A

DESCRIPTION OF THE GENUS

PINUS,

ILLUSTRATED WITH FIGURES,

DIRECTIONS RELATIVE TO THE CULTIVATION,

AND

REMARKS ON THE USES

OF

THE SEVERAL SPECIES.

BY

AYLMER BOURKE LAMBERT, ESQ. F.R.S. F.S.A.

VICE PRESIDENT OF THE LINNEAN SOCIETY.

GENERIS HUJUS SPECIES MERENTUR ATTENTIUS EXAMINARI A BOTANICIS, UT NULLUS DUBITO, QUIN PLURIMAS POSSIDEAT EUROPA
DISTINCTISSIMAS, QUÆ NUNC CONFUSÆ IGNORATÆQUE LATENT.   JACQ. IC. RAR.

LONDON:

PRINTED FOR J. WHITE, AT HORACE'S HEAD, FLEET STREET,

BY T. BENSLEY, BOLT COURT.

1803.

TO

THE RIGHT HONOURABLE

Sir JOSEPH BANKS, Baronet, K.B.

PRESIDENT OF THE ROYAL SOCIETY, &c. &c.

WHO HAS

DEDICATED THE GREATEST PART OF HIS LIFE

TO

THE PROMOTION OF NATURAL SCIENCE,

AND

RENDERING IT USEFUL TO MANKIND.

HOW EMINENTLY HE HAS SUCCEEDED ALL EUROPE IS SUFFICIENTLY INFORMED, AND HIS NAME

WILL CONTINUE AN ORNAMENT TO THE PAGES OF SCIENCE,

TILL TIME SHALL BE NO MORE.

PREFACE.

THE difficulty and obscurity of the Genus PINUS have long been remarked and regretted by Botanists; and, though so many of its species possess peculiar recommendations to the attention of horticulturists, instructions have been wanting for their better cultivation and management. It is in consequence of the growth of this tribe having been little attended to, and of authors forming their descriptions chiefly from dried and mutilated specimens, that so much confusion has prevailed. Even Linnæus himself seems to have been very partially acquainted with the changes produced by diversity of soil, and the various stages of growth; and the *Hortus Kewensis*, in which the species are certainly much better distinguished than in any other work, does not enumerate all that are now known, nor does it in every instance discriminate their characters correctly. Conceiving therefore a new arrangement of the Genus to be particularly desirable, I have devoted my attention to it for some years, and have not failed to apply to every source of information connected with the subject, having visited every plantation within many miles of the metropolis, and consulted every author of repute, with a view not only to ascertain the most accurate specific distinctions; but also to collect every fact relative to the culture and uses of every individual species. One of my objects in writing this work was to endeavour to promote the growth of deal timber in this country, which might be effected much more than at present, and would certainly prove of national importance. Neither would I overlook the ornamental part, or the improvement of the numerous plantations around the Noblemen and Gentlemen's seats in this kingdom, which at present are composed too much of one species of *Pinus*, and that not the most beautiful, the Scotch Fir. I attribute this to the different species not having been properly pointed out, a defect which is here endeavoured to be remedied. I cannot help lamenting that more has not been done in London towards the promotion of natural science in describing and publishing accounts of the numerous and interesting public museums of natural history here collected; more abundantly perhaps than in any other part of Europe. But collections are piled upon collections and altogether neglected, while new productions are sought with avidity in distant regions, and I cannot but agree with Cuvier in his excellent Eloge on the celebrated Bruguieres, that one cause of this neglect, and perhaps the chief, is the facility of procuring pleasures of all kinds in a gay and rich metropolis, added to the charms of the fascinating society in which we live; all these hold out temptations which encroach terribly on literary leisure, and only leave room for a few sacrifices to celebrity; which it must be confessed are not advanced by insulated descriptions and minute discussions.

The most remarkable gardens for the cultivation of Pines in this country are at *Pain's Hill*, which are preferable perhaps to any in Europe, both for variety of species, and excellence of growth. A considerable sum of money was formerly made by the gardener every year by selling the cones for seed. *Kew Gardens* likewise furnish many species in high perfection. Among the most striking are *Pinus palustris*, probably now the largest in England; *Pinus Cembra*, annually producing fruit; *Pinus Pumilio*, *Pinus Halepensis*, and *Pinus resinosa*.

a

There are several Pines remaining at Whitton, the seat of the late Duke of Argyll, so often referred to in the *Hortus Kewensis*. The first *Pinus Cembra* ever planted in our Island is now growing in these gardens in perfect maturity. Not less worthy of attention are two fine trees of *Pinus pendula* and *Pinus microcarpa*, bearing great quantities of cones annually. Sion, the seat of the Duke of Northumberland, furnishes many fine trees of this Genus, particularly of *Pinus resinosa* and *Pinus Tæda*. Croom, the seat of the Earl of Coventry, affords almost every species that can be procured. Here are large trees of *Pinus palustris*, *Pinus Pumilio*, *Pinus Banksiana*, &c. The perfection to which Pines arrive on a strong soil may be seen in the very extensive plantations of Lord Rivers, at Stratfieldsay, Hampshire; which, in about forty-two years have grown to a much greater size than any others I have ever seen. In the year 1799, I paid a visit with the worthy President of the Linnean Society, Dr. Smith, to the curious garden of the late Peter Collinson,[a] at Mill Hill, and was much delighted to find it nearly in the same state as it was left to his son the late Mr. Michael Collinson, who bestowed much attention upon it. We saw here three trees of *Pinus Cembra*, the finest in England, and a most flourishing *Pinus pendula*, the first that was introduced into this country. I could not help feeling regret when this delightful seat of Flora, where the owner was in frequent correspondence with Linnæus, and which contained the produce of so many distant travels, was sold by Mr. Charles Collinson, and exposed to the danger of being converted into a mere pleasure garden. It has of late however become the property of Richard Anthony Salisbury, Esq. a gentleman no less passionately attached to the study of Botany than distinguished for his accurate knowledge in that science, and in whose possession the many curious vegetable productions that are still remaining will be inviolably secured from all destruction.[b] It is proper in this place to mention how much I am indebted to the works of *Evelyn*, *Du Hamel*,[c] Hunter,[d] and Wangenheim.[e] The last in particular, which has not hitherto appeared in our own language, was found to contain so much valuable matter, that it has been quoted very largely. I ought also here to express my obligations to Dr. James Edward Smith, Dr. William George Maton, Richard Anthony Salisbury, Esq. Jonas Dryander, Esq. and William Townshend Aiton, Esq. from whose kind attention and important communications I have derived essential assistance throughout the whole progress of this work. It is my intention to follow up the present work with an illustration of the remaining Genera in the natural order of *Coniferæ*. Several drawings are already finished for that purpose of the species of *Dacrydium*, and the *Dombeya* of Lamarck, which are intended to be given to the public as soon as possible.

[a] I have lately been favoured with a sight of the remains of this celebrated naturalist's Herbarium, and have also perused many most interesting letters written by him from Mill Hill to Linnæus, and now in the hands of Dr. Smith. A great part of the Herbarium has been destroyed, and what is left contains but few specimens worthy of notice, communicated to him chiefly by John Bartram from America. There are attached to them several notes of the donor in his own hand-writing, which strongly mark the simplicity of his heart. Several of his letters are also in the Linnean Museum.

[b] In another visit to the garden at Mill Hill before it came into the possession of Mr. Salisbury, I very much lamented seeing the fine *Pinus pendula* cut down and converted into paling for part of the garden. I brought home a large piece of its timber, the grain of which, when polished, shewed itself to be equally good with that of the White Larch, *Pinus Larix*.

[c] Traité des Arbres et Arbustes, 2 Tomes.

[d] Edition of Evelyn's Sylva, 4to. 1786, and 4to. 1774.

[e] Beytrag zur tutschen holzgewächten forst wissenschaft. Folio, Gotting.

CHARACTER GENERICUS.

MONOECIA MONADELPHIA. *Linn.*   CONIFERÆ. *Juss.*

Flores amentacei monoici.

MASCULI.

CALYX. *Amenti* squama.

COROLLA nulla.

STAMINA. *Filamenta* plurima, infernè connata in columnam. *Antheræ* biloculares, supernè subcristatæ.

FŒMINEI.

CALYX. *Strobilus* constans *squamis* imbricatis, bifloris, persistentibus, induratis, ungulis *stigma* ferentibus.

COROLLA nulla.

PISTILLUM. *Germen* minimum, geminum, squamis supernè insertum. *Stylus* nullus, nisi strobili squama. *Stigma* in apice vel dorso squamæ.

SEMINA bina, oblonga, plùs minùs alata.

SYNOPSIS OF THE SPECIES.

* Foliis pluribus ex eâdem basi vaginali.

		PAGE.	TAB.
1	Pinus sylvestris	1	2.
2	- - Pumilio	5	2.
3	- - Banksiana	7	3.
4	- - Pinaster	9	- 4, 5.
5	- - Pinea	11	6, 7, 8.
6	- - maritima	13	9, 10.
7	- - Halepensis	15	11.
8	- - Massoniana	17	12.
9	- - inops	18	13.
10	- - resinosa	20	14.
11	- - variabilis	22	15.
12	- - Tæda	23	16, 17.
13	- - rigida	25	18, 19.
14	- - palustris	27	20.
15	- - longifolia	29	21.
16	- - Strobus	31	22.
17	- - Cembra	34	23, 24.
18	- - occidentalis	36	

* * Foliis solitariis ramos ambientibus.

19	- - Abies	37	25.
20	- - alba	39	26.
21	- - nigra	41	27.
22	- - rubra	43	28.
23	- - orientalis	45	29.
24	- - picea	46	30.
25	- - Balsamea	48	31.
26	- - canadensis	50	32.
27	- - taxifolia	51	33.
28	- - lanceolata	52	34.

* * * Foliis numerosis fasciculatis ex unâ basi vaginali.

29	- - Larix	53	35.
30	- - pendula	55	36.
31	- - microcarpa	56	37.
32	- - Cedrus	58	

* * * * Addenda.

| 32* | - - Dammara | 61 | 38. |

c

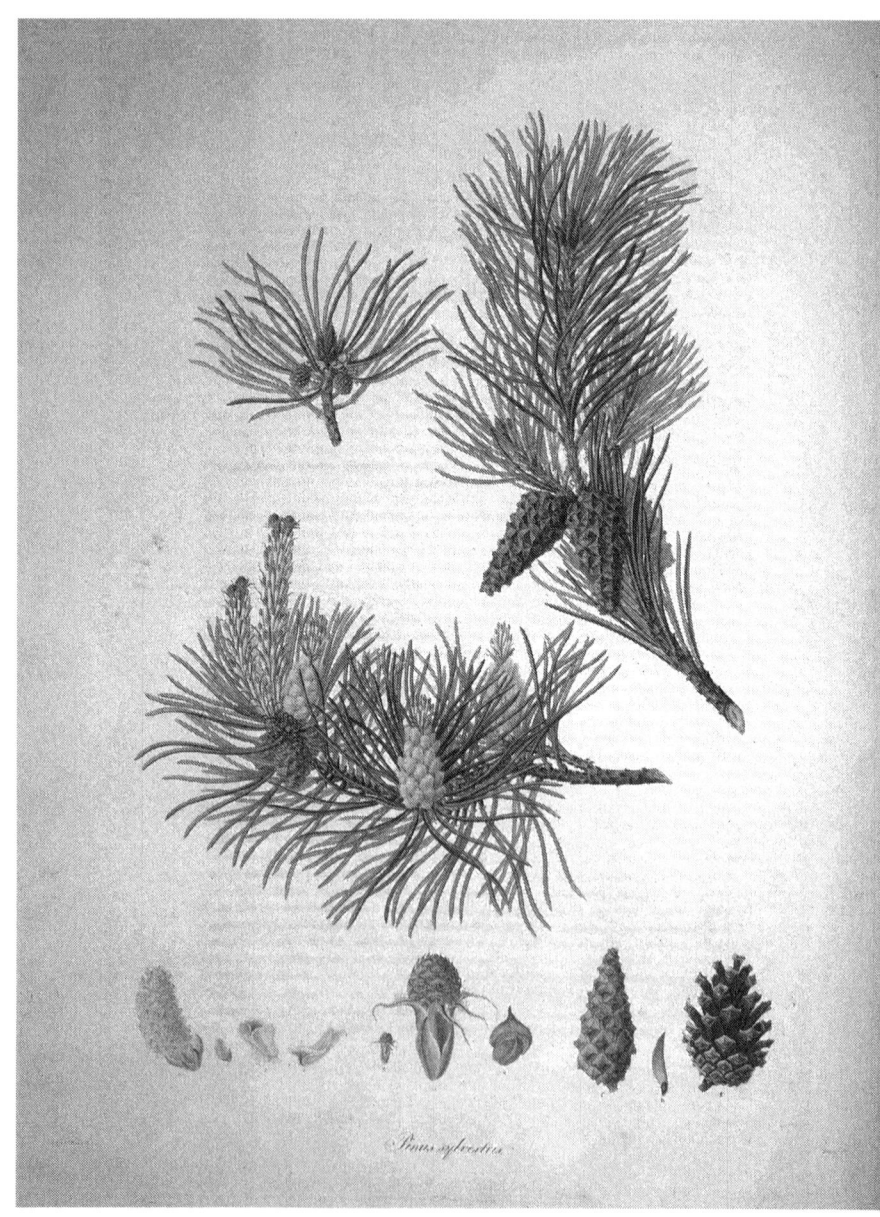

Pinus sylvestris

TAB. 1.

1. PINUS SYLVESTRIS.

SCOTCH FIR.

PINUS SYLVESTRIS, foliis geminis rigidis, strobilis junioribus pedunculatis recurvis dependentibus, antherarum cristâ exiguâ.

P. *sylvestris*, foliis geminis, rigidis; strobilis ovato-conicis, longitudine foliorum, subgeminis, basi rotundatis. *Soland. Mss.*

P. sylvestris. *Linn. Sp. Pl.* 1418. *Syst.* 860. *ed. Reich. v.* 4. 172. *Hort. Cliff.* 450. *n.* 1. *Fl. Suec. n.* 874. *Lapp. n.* 346. *Mat. Med. n.* 470. *Woodv. Med. Bot.* 570. *t.* 207. *Sm. Fl. Brit.* 1031. *Huds. Angl.* 423. *With. Arr. ed.* 3. 615. *Lightf. Scot.* 587. *Pallas. Ross. v.* 1. 5. *t.* 2. *f.* 1, i. *Scop. Carn. n.* 1196. *Pollich. Pall. n.* 913. *Gunn. Norv. n.* 337. *Villars. Dauph. v.* 3. 804. *Trew. in Nov. Act. Nat. Cur.* 3. *App.* 452. *t.* 15. *f.* 1. 3. *t.* 16. *f. n.* 25. *Mill. Illust. t.* 82. *Du Roi Harb. ed. Pott. v.* 2. 16. *Evel. Sylv. ed. Hunt. t.* 262. *Blackw. t.* 190.

P. foliis binis, convexo-concavis, conis masculis solitariis alaribus. *Hall. Helvet. n.* 1660.

P. rubra. *Mill. Dict. n.* 3.

P. sylvestris communis. *Ait. Kew. v.* 3. 366.

P. n. 29. *Gmel. Sib. v.* 1. 178.

Habitat in *Europâ boreali* sylvis glareosis.
Floret Maio.

DESCRIPTIO.

Arbor excelsa, rectiuscula, ramis obliquis. *Cortex* squamoso-deciduus. *Folia* e vaginis tubulosis, membranaceis, corrugatis, laceris, per ramulos spiraliter dispositis, geminatim prodeuntia, biunciialia, erecto-patentia, linearia, obtusa cum mucronulo cartilagineo, serrulata; suprà canaliculata; subtùs convexa, ecarinata; atro-viridia, glabra, sempervirentia. *Amenta* terminalia, pedunculata, basi bracteata: *mascula* spicata, numerosa, erecta, ovata, obtusa, flava, nuda; staminibus monadelphis, numerosis; antheris pedicellatis, cuneato-oblongis, marginatis, bilocularibus, apice cristâ membranaceâ, parvâ, sub-erosâ, auctis; *fœminea* sæpiùs terna, erecta, ovato-subrotunda, viridia, post impregnationem recurvato-pendula ac fuscescentia; squamis imbricatis, dilatatis, acuminatis; bracteis interioribus elongato-acuminatis, ciliato-dentatis. *Strobilus* secundo anno maturus, pendulus sesquiuncialis, ovato-oblongus, tessellatus, squamis angulatis, pyramidatis, retusis, tuberculosis, inermibus. *Seminum* ala verticalis, falcato-lanceolata, elongata, acuta.

B

This well known tree, though tall, seldom grows straight, and the *branches* shoot rather obliquely. The *bark* is rough and cracked. The *leaves* are short, pungent, concave on the upper surface, convex on the under, and of a pale green colour. The *male flowers* are whitish. The *pollen* is sometimes in spring carried away by the wind in such quantities, as to alarm the ignorant with the notion of its raining brimstone. The *strobili*, or cones, are small, nearly conical, and pointed; they grow to the number of two, three, or four together round the branches. While they are young, they are generally pendent, and of a purplish colour. The *squamæ*, or *scales* of the cones project in the middle, and form four distinct areæ, or compartments. The *seeds* are small, somewhat like those of *P. Abies.*

As *P. sylvestris* grows spontaneously in Scotland, Denmark, Norway, and other countries in the north of Europe, it would seem that a cold climate alone suited it, but experience proves that when it is properly reared and planted, no temperature, scarcely, impedes its growth to a considerable size.

The seeds should be procured in the following manner. The cones, which must be gathered in the winter, should be preserved until the month of June, when they must be occasionally brought forth, and exposed to the utmost heat of the sun; this will cause the scales to open, so that the seeds may easily be shaken out. They should be laid on a large carpet, or oil cloth, which will save the seeds that drop when the cones are turned, for as often as the scales on one side of them are opened, it is proper that the other should be turned to the sun to receive the same effect. These seeds will be fit to be sown in the spring following; the middle of April or May is the best time. Warm dry weather is requisite for the sowing, and a fine light mould. Beds should be made in the seminary three or four feet wide, and the seeds sown in these at a little more than a quarter of an inch in depth. The young Firs will appear in about six weeks, with the husks of the seeds on their heads, and at this period they must be carefully watched, for if the sparrows or other birds once take to them, they will destroy every plant as fast as it comes up, so fond are these creatures of the husks. In order therefore to secure the young crops, it will be proper to cover them with some good nets, and to draw over the latter strings with feathers tied across, that before they have any temptation, the birds may be frightened away, and the plants, at their first appearance, remain unnoticed by them. As soon as all the plants are come up, and have parted with their husks, the nets and strings with feathers may be taken off, for the seedlings will then be out of danger.

The following summer they will need no other care than being kept free from weeds. In the latter end of March, or the beginning of April, the second year, they should be taken out of those beds, and put into others at the distance of three or four inches from each other. When they are first removed, being one year old from the seeds, will be found to have no shoot, but are slender plants with small weak buds; and by the spring following few of them will have made a shoot, though the bud will be considerably stronger. In the spring of the third year the young Firs ought to be removed a second time, viz. into the nursery, where they should be planted about one foot asunder, and at the distance of two feet in the rows. The ensuing summer they will have grown to the height of one foot, or more. In the spring of the fourth year, if the ground designed for the plantation be ready, and there be no rabbits nor hares near the spot, they may be transplanted for the last time. If any of the animals just mentioned have the means of getting to them, it will be most advisable to defer the final removal to another year. Plantations are often wholly destroyed by hares, the winter after they are made, unless they have acquired some strength and reached the height of three or four feet. But here it ought to be remarked, that the larger the trees may be grown, the greater will be the difficulty of removing them, and when they are of a tolerable height, many will necessarily be lost after they have been transplanted. It is advisable to allow Firs, in all open situations, the distance of four feet or more, and to place them irregularly in the final place of growth. They will always flourish best when planted in turf, or where the earth has not been disturbed. (From not attending to this circumstance, it often happens that the trees become unhealthy and defective. Fruit trees, in some parts of the west of England, particularly Wiltshire, are apt to suffer in the same way, on account of having a border before them; and I have known a large garden planted three times, in consequence of this circumstance, which notwithstanding may occur only in chalky soils, as I have had no experience of other situations.) In about five or six years, the branches will have met, and begun to interfere with each other; pruning will now be necessary, but it is not to be done without great caution, and the lower branches only are to be taken away. The operation ought to be performed in September, at which time there will be no danger of the wounds bleeding too much. It should be repeated every other year, leaving all the upper branches entire; the lower should be cut close to the stem. As these trees never put forth new shoots where they are pruned, nor from below that part, so they suffer more from amputation than others. At the expiration of twelve or fourteen years, no more pruning will be necessary, for such branches as do not enjoy a free access of air, will die; but if the young trees have made good progress, it may be proper to thin them occasionally. The gardener ought to begin with those which are in the middle of the plantation, in order that they may enjoy the shelter of those which are on the outside, for a time, and then acquire strength before the whole number are exposed to the admission of a greater current of cold air. When the plantations are thinned the roots should not be torn up, lest the trees which are left standing be injured; as these roots will not shoot again, no disadvantage can arise from suffering them to decay in the earth. As the upright growth of these Firs renders their wood the more valuable, they should be left pretty close together, in order to draw each other up. Some trees will shoot to the height of twenty feet with perfect straightness. If they be left eight feet asunder each way, there will be quite sufficient room for their growth.

It is from *P. sylvestris* that the *red* Deal is obtained, as we are informed by Mr. Coxe. The *white* is from *P. Abies*, which, he says, is the most demanded, because no country produces it in such quantities as Christiania and its vicinity. One tree yields three pieces of timber eleven or twelve feet in length, and is usually sawed into three planks. Before it arrives at its greatest perfection, however, a tree must generally attain seventy or eighty years growth. *(See Northern Tour*, vol. v. p. 37. *oct. edit.)* I am indebted to Mr. Davis, of Wiltshire, for much information on the subject of Deal, as well as on the produce of other species of Pines, which the reader will find in another part of this work. It is surprising that this species is not more cultivated on waste grounds in several parts of Great Britain, as the few planted on Bagshot and Hounslow Heaths, &c. succeed so well. I have observed it thrives least on chalky land; but even there it does as well, if not better, than any other species, provided the ground be not disturbed about its roots. The *larva* of an insect, (which I suppose to be one of the under mentioned *Phalænæ*, although I have not yet had an opportunity of watching its transformation to determine the species") injures the young plantations of this tree. The *larva* introduces itself into the pith, or *medulla* of the young shoots on which it feeds, and which are soon destroyed by it. I have seen several young trees in the plantations of Henry William Portman, Esq. at Bryanston, Dorset, bearing marks of the injury alluded to; and the same circumstance I observed in the year 1801, in the large trees of this species in the plantation of William Beckford, Esq. at Fonthill, Wilts. The first volume of the *Museum Rusticum* contains a paper on this subject.

The following insects are observed to take up their abode on *P. sylvestris, viz. Phalæna sylvatica. P. catenata. P. seticornis. P. testacea. P. resinorum. Tenthredo pennacea. T. erythrocephala. Aphis Pini. Curculio Pini. C. septentrimalis. Dermestes præmorans. D. piniperda.*† (*Ips piniperda* of Marsham.) *Cimex pinetorum*, and *Acarus ruber.*

It has been supposed, from the authority of Duhamel, that *P. sylvestris* grows in St. Domingo, but the Pine sent from that island is the *P. occidentalis* of Swartz, hereafter to be mentioned.

There is a specimen in the Banksian collection, marked "*Pinus Tutarica,* from Paine's Hill," which is distinguished from *P. sylvestris,* in Dr. Solander's description, solely by the colour of the branches.

" Since writing the above I find it to be *Dermestes piniperda,* Linn. *Ips piniperda,* Marsham, Entom. Brit. 97.

† Linnæus says of this species, " *Habitat in Europæ ramulis inferioribus pini, quos perforat, exsiccat, unde naturæ hortulanus in hac arbore.*" But its depredations are not confined to the lower branches: for in the extensive plantations of Mr. Beckford, at Fonthill, so much of the medulla of the *young shoots* has been eaten through by this insect, that many fine trees have been almost destroyed; and in the year 1801, when I observed this fact, apprehensions were entertained of several others suffering the like fate,

EXPLANATION OF TAB. 1.

The Plate exhibits branches of *Pinus sylvestris*, which, like the figures in all the following Plates, where the contrary is not expressed, are of the natural size.

In the detail of the fructification such parts as are of the natural size are distinguished by small letters, the magnified ones being always marked with capitals.

a, A. Male catkin with its bracteæ.
B, B. Anthera.
C, C. Crest of the Anthera.
d, D. Female catkin with its bracteæ.
E. A separate scale.
f. A ripe cone.
g. The same expanded by drought.
h. Seed with its wing.

Tab. II.

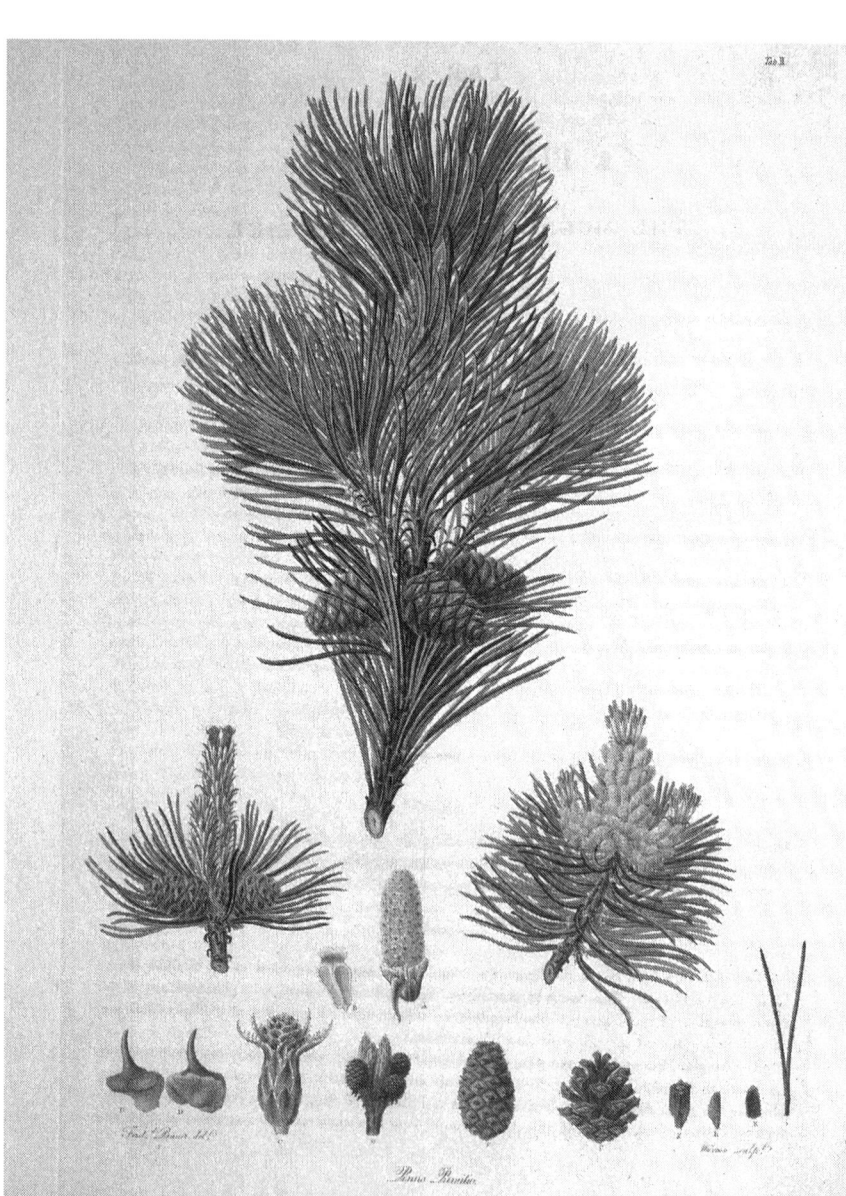

Pinus Pinaster

TAB. 2.

2. PINUS PUMILIO.

THE MUGHO, OR MOUNTAIN PINE.

KRUMHOLZ. *Germ.*

Pinus Pumilio foliis geminis abbreviatis strictis, strobilis ovatis obtusis minimis: junioribus sessilibus
erectis.

P. sylvestris montana *γ. Ait. Kew. v.* 3. 366.
P. sylves. pumilis *γ. Neal. Cat. Hort. Blackb.* 50.
P. Mughus. *Du Roi Harbk. ed. Pott. v.* 2. 41. *Scop. Carn. n.* 1195. *Wilden. Berlin. Baumz.* 206.
P. conis erectis. *Tournef. Inst.* 586. *Scheuchz. It.* 460. *Duhamel. Arb. v.* 2. 126. *n.* 13.
P. humilis, julo purpurascente. *Tournef. Inst.* 586. *Duhamel. Arb. v.* 2. 126. *n.* 12.
P. sudeticus seu carpaticus. *Ungarisch. Mag.* 3 *ter band.* 38.
Pinaster conis erectis. *Bauh. Pin.* 492.
P. Pumilio ex monte Arbâ Bavariæ. *Camer. Hort.* 127.
P. Pumilio montanus. *Park.* 1537. *f.* 8.
P. Pumilio. *Clus. Pan.* 15. *Hœncke Beob.* 68. *Hall. Helvet. n.* 1660. *γ.*
P. quartus austriacus. *Clus. Hist. v.* 1. 32.

Habitat in montosis Europæ australis.
Floret Junio.

DESCRIPTIO.

Minima in hoc genere. Præcedente longe humilior, vix septempedalis, sarmentis repentibus, ramu-
lisque radicantibus. *Folia* minora. *Antherarum* crista ampliata, biloba. *Amenta fœminea* nunquàm
arcuato-recurva. *Strobili* ovati, obtusi, duplo quàm in præcedente minores.

The specimen from which the figure was taken, was obligingly presented to me by John Black-
barn, Esq. of Orford. The tree was planted by that gentleman's father, who possessed one of the
finest collections of exotic plants in the kingdom, an account of which may be seen in Neal's Catalogue,
and which is still kept up with great care and attention.

The Mugho Pine grows on the tops of the highest mountains, where scarcely any other tree is to
be found, and it often covers with its thick and almost impenetrable branches a very extensive tract.
Hæncke has given the most complete description of it that I can find, and this is copied by our countryman Townson, who observed this species to be very exuberant on the mountains of the more

D

northern parts of Hungary. The roots generally run, it seems, in an oblique or subhorizontal direction; they are long, thick, and hard, clothed above ground with a brownish bark, shooting often to a considerable extent quite bare of earth. The branches proceed either immediately from the root, or from a low radicating trunk, scattered, long, and pliable. They are commonly about four or five feet in height, but in some instances will exceed that of a man by one foot or more. On the upper part the branches are extremely thick, and covered with a strong ash-coloured bark, which is rendered very rough and uneven by the tubercles of the fallen leaves. The smallest branches are very short as well as thick, bent in at the base, and naked to a certain height, but at the upper part they are profusely leaved and folded within each other. The leaves spring from a dry, jagged, brownish sheath, and are of a woody texture, being firm and tough. They are slightly incurvated, often twisted, and obtuse. The under surface is flat, or but slightly concave, the upper convex, the margins are minutely serrated. They are smooth, shining, faintly striated, and of a green colour, approaching to yellow at the points. Their length is from one to one inch and a half, and the breadth scarcely one fourth part of a line. The *male flowers* are terminal, and grow several in a bunch. The *female* lateral, sessile, invariably erect, sometimes single, sometimes collected into a bunch to the number of ten or twelve, ovate or subglobose, and resemble very much those of *P. Larix*, both in size and shape; in colour they are brownish, or inclined to purple. The *squamæ*, or scales, are imbricated, in their more advanced state, often open and without the *apex* that appears in the earlier ones. There is a gibbosity outwards; and on the inner side, somewhat of a concavity is observable. This tree though of humble growth, when planted on a lawn, assumes a handsome and ornamental appearance. It was supposed to be a variety of *P. sylvestris*, but I had made a distinct species of it before I saw Hænke's description. What distinguishes it particularly from the latter is, the young cones which grow erect and sessile until they are above one year old, when they become horizontal; and they can scarcely be said to be pendent, even when they are full grown; whereas those of *P. sylvestris* have long peduncles, and become pendent soon after they are impregnated with the pollen. The cones of *P. Pumilio* are of a looser texture, and but slightly attached to the tree. When the branches of this tree are broken, a transparent resin of a very fragrant smell exudes, and this is collected and sold in the form of a native Balsam. A sort of empyreumatic æthereal oil is obtained by distillation from the burned branches, and sold in Hungary, under the name of Krumholz oil.

There is a specimen in the Herbarium of Sir Joseph Banks, marked in Miller's own hand-writing *Pinus Tatarica*, which without doubt belongs to *Pinus Pumilio*.

Pinus Mughus, Jacq. Ic. rar. tab. 193, does not belong to *P. Pumilio*, but appears to be only a variety of *P. sylvestris* from a specimen I examined in the Herbarium at Oxford.

EXPLANATION OF TAB. 2.

A.   Male Catkin.

B.   Anthera.

C.   Female Catkin.

D, D. Separate scales.

e.   Young cones in their natural situation.

f. f. Ripe cones.

g.   A separate scale.

h.   Seed.

i.   Leaves with their sheath.

K.   Point of a leaf magnified.

Tab. 18.

Pinus Banksiana

TAB. 3.

3. PINUS BANKSIANA.

LABRADOR PINE.

Pinus Banksiana foliis geminis divaricatis obliquis, strobilis recurvis tortis, antherarum cristâ dilatatâ.

P. sylvestris divaricata *l. Ait. Kew. v. 3. 366.*
P. canadensis bifolia, foliis curtis et falcatis, conis mediis incurvis. *Duhamel. Arb. v. 2. 126. n. 10.?*

Habitat in Americâ septentrionali.
Floret Maio.

DESCRIPTIO.

Arbor ramosissima, patula, ramis longissimis. *Folia* uncialia, falcata. *Amenta mascula* cylindracea; antheris sessilibus, cristâ reniformi, emarginatâ, crenatâ, utrinque prominulâ. *Strobili* bini vel terni, sessiles, magnitudine *P. sylvestris* sed graciliores, pallidiores, flavescentes, acuminati, et insigniter incurvato-torti.

The specimen represented in the plate was taken from a remarkably fine tree growing at 'Pain's Hill, Surrey. The branches of this tree bore more fruit than any species I have seen. The cones were not more than five or six inches distant from one another in scarcely any part of the tree, and they were growing two or three together. Many of the young shoots were covered with resin, the odour of which was inconceivably fragrant. It flowered, I was informed, earlier in the spring than any other Pine. It is most partial to a sandy soil. The *branches* shoot very thickly almost the whole of its height, and consequently render the timber too knotty to be made into good masts, though it is very pliable, and contains a great quantity of resin. The *leaves* do not differ much from those of *P. sylvestris*, except that they are curved and divaricated, the pairs touching each other at their extremities so as to form a sort of ring. The *cones* are curved in a similar manner, having the appearance of horns springing from the branches; they are of nearly the same thickness as those of the Scotch Fir, but rather longer. At present *P. Banksiana* is very rare in England; I know only three of any size, one of which is at Pain's Hill, and this is certainly the finest; one at Kew, and the other at Croome, the seat of the Earl of Coventry. It is surprising that this species should ever have been supposed to be a variety of *P. sylvestris*, the one being an American, the other an European tree. I am not acquainted with any author who has noticed it, except Mr. Aiton in his *Hortus Kewensis*. Whether Duhamel's species above quoted be the same, may be questioned; but as he mentions the cones to be remarkably contorted,

---

' This beautiful spot was the seat of the late Hon. Charles Hamilton. It is now in the occupation of —— Moffat, Esq. Linnæus the son visited these gardens in company with Sir Joseph Banks, and expressed himself highly gratified in viewing their productions.

E

I have given his synonym with a doubt. By whom this species was first introduced into England I have not yet learned. Mr. Forsyth, of Kensington Gardens, received a tree of it some years ago from a person who had been sent into the interior parts of America by the late Dr. Fothergill; this probably was the first that found its way into England.

As I am entirely obliged to Sir Joseph Banks for the first knowledge of this species, I have given it his name.

EXPLANATION OF TAB. 8.

a. Leaves with their sheath.
B. Male Catkin.
C. Anthera.
d. Female Catkin.
e. Ripe Cone.
f. Scale of the Cone.
g. The same seen on its inside.
h. Seeds.
i. A Seed without its wing.

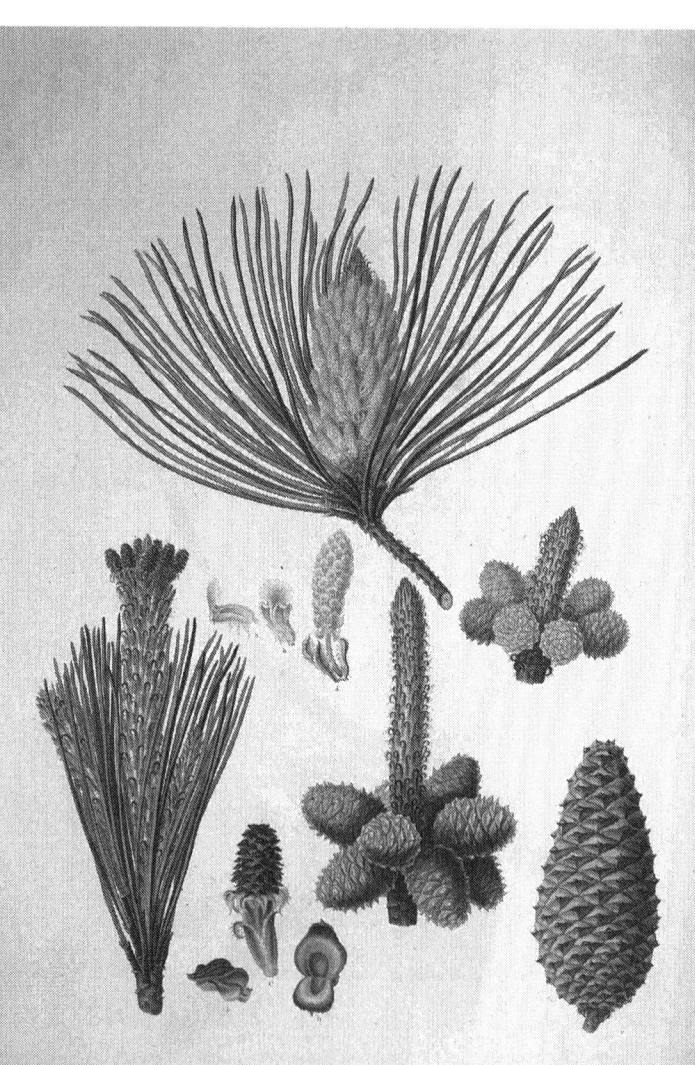

Pinus Pinaster

Pinus Pinaster.

TAB. 4 & 5.

4. PINUS PINASTER.

THE PINASTER, OR CLUSTER PINE.

Pinus Pinaster, foliis geminis elongatis, strobilis verticillatis confertis ovatis sessilibus pendulis, antherarum cristâ rotundatâ.

P. *Pinaster*, foliis geminis, margine subasperis, conis oblongo-conicis, folio brevioribus, basi attenuatis, squamis echinatis. *Soland. MSS. Ait. Kew. v. 3. 367.*

P. sylvestris γ. *Linn. Syst. Reich. v. 4. 172.*

P. maritima altera. *Duhamel. Arb. v. 2. 125. n. 4. t. 29. Du Roi. Harb. ed. Pott. v. 2. 59.*

Habitat in Europæ australis maritimis.
Floret Maio.

DESCRIPTIO.

Arbor excelsa, ramis patentibus, subfastigiatis. *Folia* quadriuncialia, recta, canaliculata, pungentia, lævia; vaginis ferè uncialibus. *Amenta mascula* pedicellata, elliptico-oblonga; antheris subpedicellatis, cristâ rotundatâ, indivisâ, dentato-lacerâ, latitudine antherarum. *Bracteæ* omnes setaceo-dentatæ. *Strobili* verticillati, numerosi, sessiles, demùm penduli, ovati, recti, magni, 5-7 unciales, squamis submuricatis. *Semina* parva, alâ elongatâ, reusâ.

P. *Pinaster* is frequent in English plantations, and grows to a great height and size, being very shewy, and bearing large shining cones, it is extremely ornamental, except in its more advanced age, when the branches become naked and very unsightly. The wood is soft, and therefore not so valuable as that of many other trees of this genus. On the mountains of Switzerland the native forests are seldom suffered to stand; being usually either cut into shingle for covering the roofs of houses, or employed for the extraction of pitch. In the south of France the young trees are made into stakes for supporting the vines. The *branches* grow at a wider distance from one another than those of P. *sylvestris* and more horizontally. The *leaves* are much larger, thicker, and longer, and have a broad surface, with a furrow running longitudinally. The *cones* are five or six inches long, and grow in very large clusters. Mr. Tucker, of Devonshire, I have been informed, has a tree that once bore as many as eighty in one bunch. The *seeds* are oblong, a little flattened at the sides, and have narrow wings. The largest trees of this species that I have seen are growing at Pain's Hill. The first *Pinus Pinaster* planted in England, was in Bishop Compton's garden at Fulham, and is still growing there in a healthy state.

F

EXPLANATION OF TAB. 4 & 5.

TAB. 4, was taken from a fine tree in the Royal Gardens at Kew.

A.   Male Catkin.
B, B. Anthers.
C.   Female Catkin.
D, D. Scales.
e.   Ripe Cone.

TAB. 5, is copied from a drawing by Ehret, in the possession of Sir Joseph Banks, by whom it was purchased, together with drawings of several other species of *Pinus*, at the sale of the late Mr. Moore's effects in Shropshire.

a.   Scale of a Cone.
b.   Inner side of the same.
c, c. Seeds.

Tab. VI.

Pinus Larix.

Pinus Pinaster

Tab. VIII.

5. PINUS PINEA.

THE STONE PINE.

PINUS PINEA, foliis geminis, strobilis ovatis maximis, seminum alis abbreviatissimis, antherarum cristâ dentato-lacerâ.

P. *Pinea*, foliis geminis, primordialibus ciliatis, conis ovatis, obtusis, subinermibus, folio longioribus, nucibus duris. *Soland. MSS. Ait. Kew. v. 3. 368. Wilden. Berlin. Baumz.* 209.

P. *Pinea*, foliis geminis: primordialibus solitariis ciliatis. *Linn. Sp. Pl.* 1419. *Syst. ed. Reich. v. 4.* 173. *Hort. Cliff.* 450. *n.* 2. *Hort. Ups.* 288. *Mat. Med. n.* 471. *Gouan. Hort.* 494. *Mill. Dict. n.* 2. *Scop. Carn. n.* 1197. *Regn. Bot. Evel. Sylv. ed. Hunter.* 266. *fol. Villars. Dauph. v.* 3. 806. *Allion. Ped. v.* 2. 177. *Vitm. Sp. Pl. v.* 5. 344.

P. *Pinea*, foliis geminis, conis pyramidatis, splendentibus, squamis oblongis, obtusis, nucibus ovatis, alâ membranaceâ destitutis. *Du Roi. Harb. ed. Pott.* 2. 52.

P. sativa. *Bauh. Pin.* 491. *Blackw. t.* 189. *Duhamel. Arb. v.* 2. 125. *n.* 1. *t.* 27.

P. ossiculis duris, foliis longis. *Bauh. Hist. v.* 1. *p.* 2. 248.

P. domestica. *Matth. Com.* 87. *Tabern. Ic.* 936.

Zinbellaum. *Linn. Pflanzen Syst.* 2. 351.

Le Pin. *Regnault. Bot. Ic.*

Habitat in Europâ Australi, Africâ Septentrionali.

Floret Maio.

Distinguitur foliis longis, geminis; strobilo ovato, obtuso, maximo; squamis crassis, apice latis, obtusis; nuce oblongo, magno, tereti. *Desfont. Fl. Atlant. v.* 2. 352.

DESCRIPTIO.

Habitus P. Pinastri, sed *folia* parùm minora, vaginis brevioribus. *Amenta mascula* vix pedicellata, antherarum cristâ reniformi, subbilobâ, dentato-lacerâ: *fœminea* globosa, erecta, squamis deflexis, suprà carinatis. *Bracteæ* integræ. *Strobili* solitarii vel oppositi, patentes, subsessiles, ovati, obtusi, maximi, crassi, tuberculosi nec muricati. *Semina* omnium in hoc genere maxima, ossea, obovata, alâ brevissimâ, retusâ.

P. *Pinea* grows to a considerable height, and is generally pretty straight. The *leaves* are about five or six inches long, thick, of a fine green colour, inserted in pairs in a common sheath; they are rounded on one side, but that on which they touch each other is flat. The *Male Flowers* are in large red bunches, and those of both sexes sometimes appear at the extremity of the same branch. The *cones* are very large, nearly ovate, and often four inches and a half in length. They consist of very hard *scales*, each having a sort of knob in the middle. The *nuts* are of a large size, and very hard. They contain kernels which have the sweetness of Almonds. A pleasant oil is obtained by expression. The wood of this tree is tolerably white and resinous, and good boards may often be made of it; but on the whole it is not so valuable as that of many other species of Pines. It is cultivated principally on account of the foliage, and the goodness of the fruit, which last is now become an article of sale in England, and may be found in several of the London fruit-shops. There is a variety, however, as I am informed by Mr. Correa, with respect to these nuts, known in Portugal, by the name of *Pinhao molar*, and at Naples called *Pignuolo molese*. This sort is quite soft. In some countries, particularly in Portugal, the cones are sometimes used for fuel.[a]

EXPLANATION OF TAB. 6, 7, & 8.

TAB. 6 & 7 were taken from a very flourishing tree in the garden of Henry Cavendish, Esq. at Clapham.

TAB. 6, a. Leaves with their sheath.
  B. Point of a leaf.
  C. Male Catkin.
  D. Anthera.
  e. Female Catkins of one year's growth.
  f. Cones of two years' growth.

TAB. 7, a. Cones of three years' growth.
  b, b. Scales of the same.
  c. Seed.
  d. The same deprived of its wing.
  e. Hard shell of the seed.
  f. Kernel.
  g. Ripe cone of four years' growth.

TAB. 8 was taken from a cone purchased in a shop to shew the greater perfection of those brought from abroad, which are in general distinguished by a small protuberance at the top.
  a, a. Scales shewing the natural situation of the seeds.
  b. Wing of the seed.
  c. Seed without its wing.
  d. Kernel in its shell.
  e, e. The same separate.
  f. Section of the kernel shewing the embryo.
  G. Embryo magnified.

[a] The fruit of Pinus Pinea is four years coming to maturity from its first formation. I have represented each year of its growth in Plate 6 and 7.

Pinus maritima

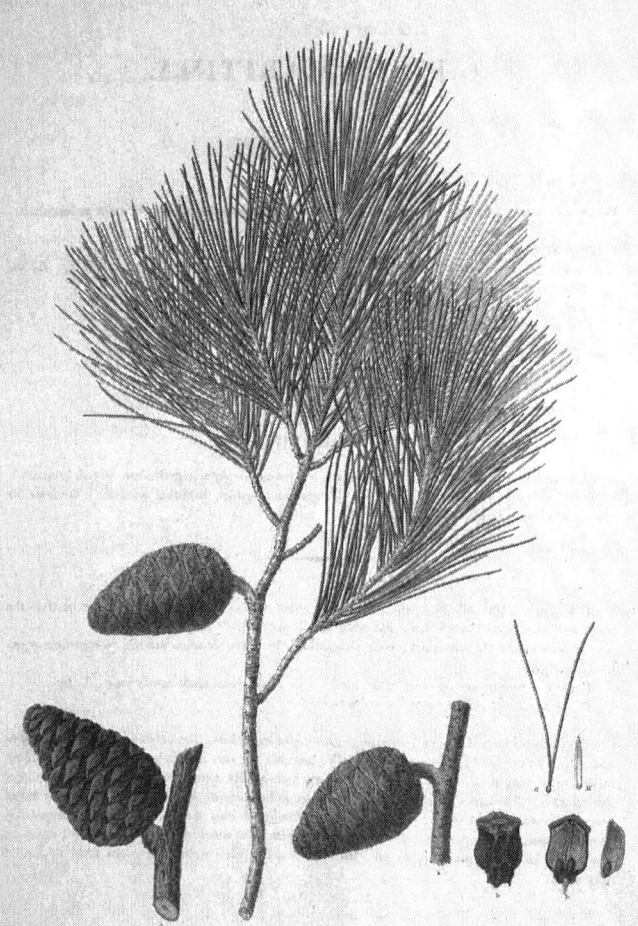

Pl. X.

Pinus maritima.

TAB. 9 & 10.

6. PINUS MARITIMA.

MARITIME PINE.

Pinus maritima, foliis geminis tenuissimis, strobilis ovato-conicis glaberrimis solitariis pedunculatis.

P. sylvestris maritima *.*   *Ait. Kew. v. 3. 366.*
P. foliis binis in summitate ramorum fasciculatim collectis.   *Duhamel. Arb. v. 2. 125. n. 3.   Du Roi Harb. ed. Pott. 2. 59.?*

Habitat in Europæ Australis maritimis.
Floret Junio.

DESCRIPTIO.

Arbor 20-pedalis, ramosissima.   *Folia* biuncialia, vel parùm longiora, angustissime, vaginâ brevissimâ.   *Strobili* solitarii, pedunculati, cernui, ovati, superficie, equales, lævissimi ac nitidi.   *Seminum* ala magna, securiformis.

The figure in the 10th Plate representing the above species was drawn from a specimen in the Sherardian Herbarium, to which the following note is annexed.

" P. maritima foliis tenuissimis, conis albicantibus, brevibus, deorsùm reflexis, in superficie æqualibus."   *Michel.*

Pinastri alterum genus parvum, in maritimis, foliis capillamenti modo tenuissimis.   *C. Iso'.*

P. maritima, conis cinereis, planis.   *Phytopin.*

This tree, so far as I can judge from one growing at Sion House, the only one I have been able to find, grows to the height of about 20 feet. The branches are very numerous, and bear long, filiform leaves, resembling those of *P. halepensis*, which are more closely connected towards the extremities of the branches. The cones are of nearly the same size as in *P. rigida*, but rather smaller. They are so remarkably smooth and glossy, that they at once distinguish this species. Those which appear on Sherard's specimen hang downwards; but those, which I obtained at Sion House, point upwards; one of the latter is represented in plate 10. In shedding their seeds, the cones seem to expand very little.

H

EXPLANATION OF TAB. 9 & 10.

Branch in flower from the Royal Gardens, at Kew, first observed June 10, 1802, since the above was written.

Tab. 9. A.   Male Catkin, magnified.
B, B.   Antheræ.
C, C.   Points of leaves magnified.

Tab. 10. a.   Cone from the Sherardian Herbarium.
b, b.   Scales of the same, with the seeds.
c.   Separate seed.
d.   Leaves.
E.   Point of a leaf.
f.   Cone from a tree in Sion Gardens.

Tab. XI.

Pinus halepensis.

TAB. 11.

7. PINUS HALEPENSIS.

ALEPPO PINE.

PINUS HALEPENSIS, foliis geminis tenuissimis, strobilis ovato-oblongis reflexis lævibus solitariis pedunculatis.

P. *halepensis*, foliis geminis, conis ovato-conicis, basi rotundatis, folio subbrevioribus: squamis obtusis.
Ait. Kew. v. 3. 367.

P. *halepensis*, foliis geminis, filiformibus; strobilo ovato-oblongo, deorsum inflexo; squamis lævibus
obtusis. *Desfont. Fl. Atlant. v.* 2. 352.

P. hierosolymitana, prælongis et tenuissimis viridibus foliis. *Duhamel Arb. v.* 2. 126. *n.* 14.

P. halepensis, foliis geminis, tenuissimis; conis obtusis, ramis patulis. *Mill. Dict. n.* 8. *Ic.* 139. *t.* 208.

Habitat in Europæ Australis, & Asiæ, maritimis.
Floret Maio.

DESCRIPTIO.

Arbor humilis. *Folia* tenuissima, fere præcedentis. *Amenta mascula* vix pedicellata; antherarum
cristâ reniformi, antheris latiori, dentato-repandâ; *fœminea* solitaria, pedunculata, mox reflexa, globosa,
squamis deflexis. *Strobili* penduli, ovato-oblongi, tuberculosi nec læves, neque muricati. *Semen*
parvum, alâ securiformi, magnâ.

" *Arbor* 7—10 metr. *Trunci* diameter in adultis vix 3 decimetr. *Rami* expansi. *Folia* gemina,
lævia, rigidula, ferè filiformia, 8 centimetr. longa, lætè viridantia, nec glauca ut in P. *sylvestris Linn.*
Strobilus deorsum inflexus, ovato-oblongus, subacutus, 2—3 centimetr. crassus, 5—8 longus. *Squamæ*
læves, obtusæ, apice duplò triplòve latiores quàm in P. sylvestris." *Desfont.*

THE Figure was taken from a specimen in the Sherardian Herbarium, which has the following
inscription annexed to it.

" P. maritima, tuberculosis conis, spadiceis, lucidis. *Phytopin.*

P. maritima prima. *Matth. Comm.*

Nomine seq. misit Micheli, P. maritima foliis tenuissimis, conis longis, angustis, nitidis. "

This species in England is more like a shrub than a tree, and never grows to any great height in
its native country. It is often greatly injured by the cold of our winters, and sometimes killed by

intense frost. In 1789 all the Aleppo Pines that were in the Paris gardens perished, in consequence of the great cold of that year. This tree has been found in a wild state, however, in the southern parts of France, as we are informed by M. Desfontaines. That botanist observed it growing on the coast about Frejus. *P. halepensis* is very readily distinguished from other species, and after what has been quoted from authors above, it will only be necessary to remark that the narrowness of the leaves, the very broad *apices* of the scales, and the obliquity of the fruit-stalks are the principal characteristics. Early in the spring a palish resin flows in large quantities from the fissures in the bark, and it will sometimes cover the branches, and even the trunk completely. According to Miller this species was first introduced into England by Consul Cox in the year 1732. There are a few Aleppo Pines three or four feet high growing in Burchell's nursery, at Fulham; they were raised from seed sent by Mr. Williams from France. A very flourishing tree is to be seen at Kew. This species is very scarce at present in England.

EXPLANATION OF TAB. 11.

a, A. Male Catkin.
B, B. Anthera.
c, c. Ripe Cone.
d. Seed with its wing.

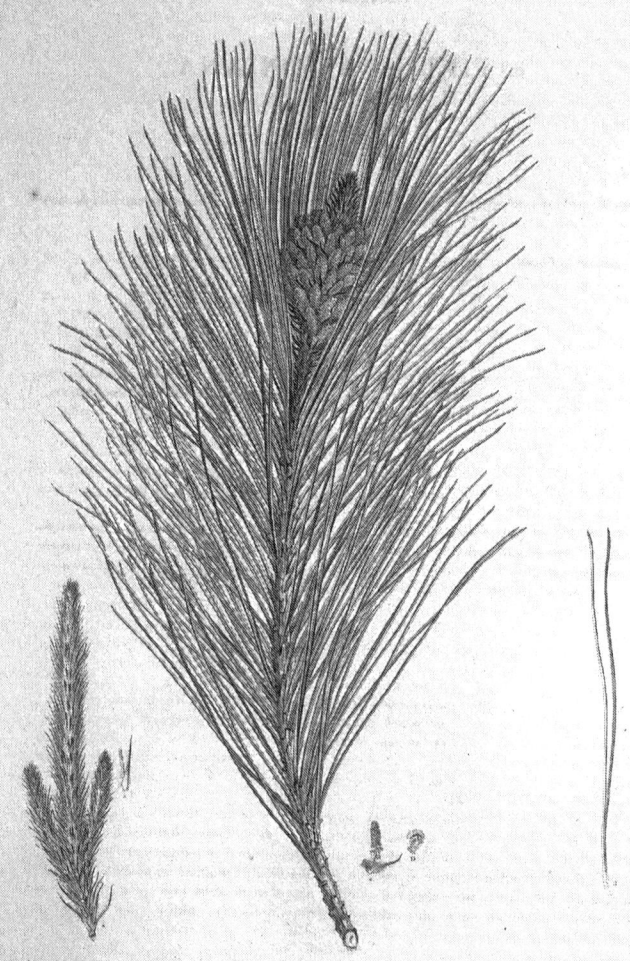

Pinus halepensis

TAB. 12.

8. PINUS MASSONIANA.

INDIAN PINE.

PINUS MASSONIANA, foliis geminis tenuissimis longissimis; vaginâ abbreviatâ, antherarum cristâ dentato-lacerâ.

Habitat in *China*.

DESCRIPTIO.

Arbor stipulis ciliato-pilosis, vaginis filamentoso-laceris. *Folia* 3—4 uncialia, angusta, canaliculata, margine scabra. *Amenta mascula* pedicellata. *Antherarum* crista plana, reniformis, dentato-lacera.

THE specimen represented in this plate is in the Banksian Herbarium, having been brought by Mr. Francis Masson from the Cape of Good Hope, where it was raised from seeds which had been sent from China.

This is a very distinct species from Dr. Roxburgh's *P. longifolia*, specimens of which, and his description, I have lately obtained. I have not been able to obtain any fruit of *P. Massoniana*, nor any further information.

EXPLANATION OF TAB. 12.

a. Stipulæ.
b. Leaves with their short sheath.
c. Male Catkin.
D. Anthera.

TAB. 13.

9. PINUS INOPS.

JERSEY PINE.

PINUS INOPS, foliis geminis, strobilis recurvis oblongo-conicis longitudine foliorum; aculeis squamarum subulatis rectis.

P. *inops*, foliis geminis; conis oblongo-conicis, longitudine foliorum, solitariis, basi rotundatis; squamis echinatis. *Soland. MSS. Ait. Kew. v.* 3. 367. *Willden. Berl. Baumz.* 208.

P. *virginiana*, foliis geminis, conis oblongis, incurvis, aculeis squamarum rectis. *Du Roi. Harbk. ed. Pott. v.* 2. 47.

P. *virginiana*, foliis geminis, brevioribus, conis parvis, squamis acutis. *Mill. Dict. n.* 9. *Wangenh. Beytr.* 74.

Habitat in Americâ Septentrionali.
Floret Maio.

———⬥———

DESCRIPTIO.

Arbor tortuosa, ramosa, 40-pedalis. *Folia* biuncialia, canaliculata, subtùs convexa. *Amenta* cylindracea; antherarum crista reniformis, dentato-lacera, antheris latior: *fœmineorum* squamæ acuminato-subulatæ, patentes, subrecurvæ. *Strobili* breviùs pedunculati, solitarii vel bini, recurvato-penduli, ovato-cylindracei, acutiusculi, magnitudine varii, squamis mucronato-spinosis, spinis rectis, vel parùm recurvis.

———⬥———

THE specimen from which the figure is taken was procured at Pain's Hill.

This species, Wangenheim remarks, is found principally in the interior parts of North America, upon mountains and hills, in a dry soil, composed of sand and pebbles. Its stem is seldom very straight; the length of it is usually from fifteen to twenty feet, and the circumference about one foot and a half. It divides into several branches growing at some distance from one another, but not in a very orderly manner. The entire height is commonly almost forty feet. The *bark* is deeply cracked, of a brownish colour, and the *wood* is of a reddish yellow. In regard to durability, the wood is apt to fail, but it abounds with resin, which, working through the fissures in the bark, gives the branches the appearance of being candied over with sugar, so that the valuable part of the tree is its pitch and tar. The *branches* indeed are tough, and pliable, and may therefore be useful for hoops, baskets, &c. Most of the Pine tribe are very brittle in their texture, but in this species the wood has almost pliability enough to be tied in a knot. The *leaves* are two inches long, pointed, rounded on the under surface,

Tab XII

Pinus inops.

furrowed on the upper; and of a dark green colour; never more than two occupy one sheath. In New York, under the forty-first degree of north latitude this Pine blossoms at the beginning of May, and the seed is ripe in November. The cones diminish to a point. Their length is not much more than two inches, and the thickness at the base, is about an inch and a half. The scales are of a hard, woody texture, and of a yellowish brown colour, and a sharp woody spine projects from each. These spines are never curved, but invariably straight.

Kalm mentions a curious fact relative to this particular species. In the heat of summer, he says, the cattle resort to its shade in preference to that of any other tree, the foliage of which may be much thicker; he himself saw them studiously singling out *P. inops*, in the wilds of America. *See Travels in North America, Foster's ed. v.* 1. 335. Hence it seems that the resinous effluvia of this tree is peculiarly agreeable to such animals.

One of the finest Jersey Pines that I have seen is in a plantation by the road side at Mr. Horne's, near Southampton.

EXPLANATION OF TAB. 13.

a. A. Male Catkin.
B, B. Antheræ.
c, C. Female Catkin.
D. Point of one of the scales.
e. Ripe cone.
f. One of its scales.
g. Seeds.
h. Leaves with their sheath.
I. Point of a leaf.

TAB. 14.

10. PINUS RESINOSA.

PITCH PINE.

PINUS RESINOSA, foliis geminis, strobilis ovato-conicis sessilibus ternis; squamis medio dilatatis inermibus.

P. *resinosa*, foliis geminis, conis ovato-conicis, basi rotundatis, solitariis, folio dimidio brevioribus, squamis inermibus. *Soland. MSS. Ait. Kew.* v. 3. 367.

P. Canadensis bifolia, conis mediis ovatis. *Duhamel. Arb.* v. 2. 125. n. 8.

Habitat in Americâ Septentrionali.
Floret Maio.

DESCRIPTIO.

Arbor mediocris. *Folia* 4-5 uncialia, margine apiceque scabriuscula, subtùs subcarinata. *Vaginæ* ferè unciales, demùm corrugatæ. *Amenta mascula* formosa, purpurascentia; antherarum crista convexa, reniformis, dentato-lacera, antheris angustior: *fœminea* ovata, obtusa. *Strobili* bini vel terni, patentes, ovati, obtusi, tuberculosi, inermes, squamis medio dilatatis.

THE figure was taken from a specimen procured at Pain's Hill, where, as also at Caen Wood (in a small island), this species grows in a very flourishing state. But the greatest number of these trees, are at Sion-House, the seat of the Duke of Northumberland, where the first that grew in England were raised. Though *P. resinosa* is rarely to be found in our Plantations, yet it is certainly entitled to more general cultivation, being of a very elegant appearance, and remarkable for the fragrance of its resin, which is very abundant. From the size of those trees which I have seen, I conclude that this species cannot produce valuable timber. It flourishes best, like many others of this tribe, in a moist situation, and a light, sandy soil. One of the most distinctive characters of *P. resinosa* is the uncommon length of the *vaginæ* or sheaths of the leaves, which resemble in their form those of *P. sylvestris*, but are generally much longer. The *inflorescentia* assumes a dark red colour, and is larger than in most other Pines. The *Cones* are shorter than those of *P. sylvestris*, and more obtuse. The bases of the *squamæ* are very blunt, and internally they exhibit the same colour as the Flowers.

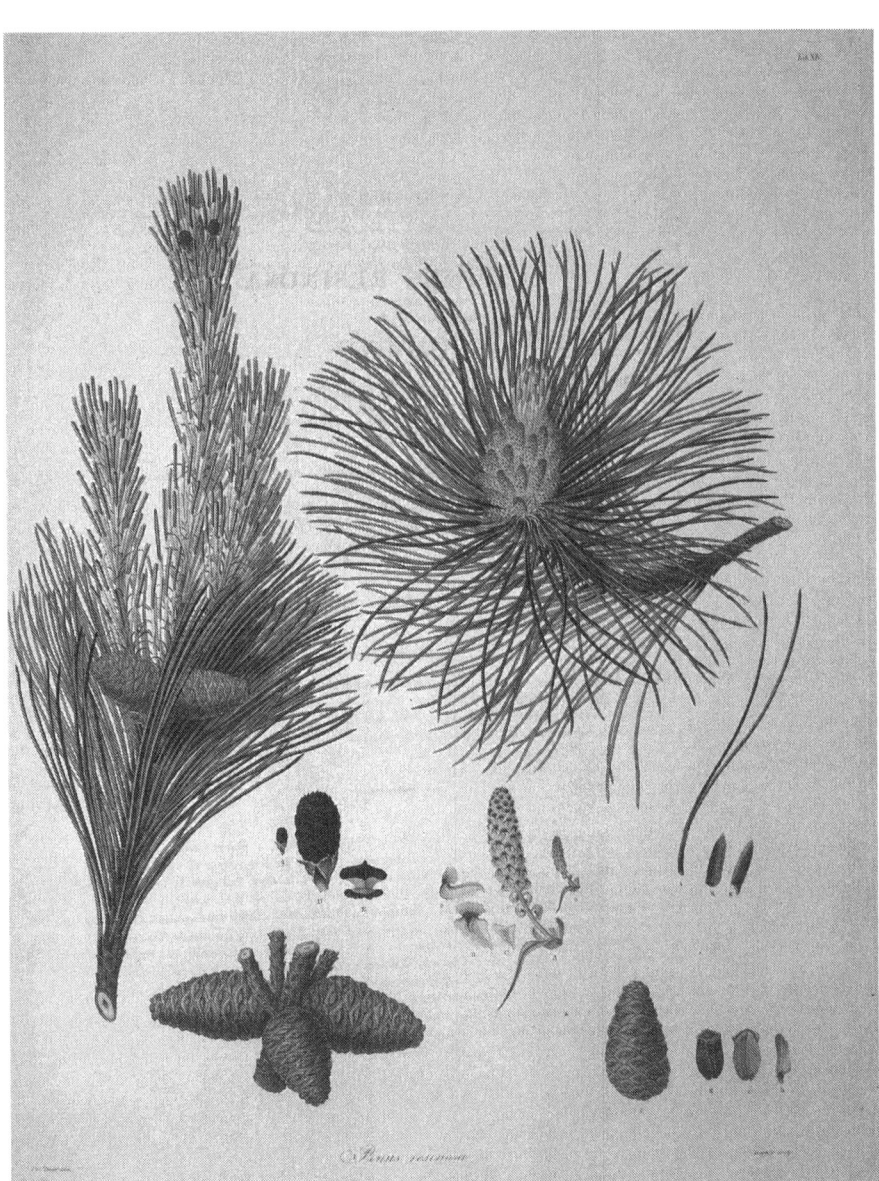

Tab. XV.

Pinus resinosa

EXPLANATION OF TAB. 14.

a, A. Male Catkin.
B, B. Antheræ.
C. Transverse section of the Anthera.
d, D. Female Catkin.
E. Scale of the same.
f. Ripe cone.
g, g. Its scales.
h. A Seed.
i. Leaves and sheath.
K, K. Points of leaves.

TAB. 15.

11. PINUS VARIABILIS.

VARIABLE-LEAVED BASTARD PINE.

PINUS VARIABILIS, foliis binatis ternatisque, strobilis ovato-conicis subsolitariis; squamarum aculeis incurvis.

P. *Tæda variabilis* ᵧ. *Ait. Kew. v.* 3. 368.
P. *echinata*, prælongis foliis tenuioribus, cono echinato gracili. *Mill. Dict. n.* 12. *Wangenh. Beyt.* 74.
P. echinata. *Marshall. Arb. Amer.* 100.
P. *echinata*, foliis geminis et ternis, conis oblongis incurvis; aculeis squamarum reflexis. *Du Roi. Harbck. ed. Pott. v.* 2. 51.

Habitat in America Septentrionali.
Floret Maio.

DESCRIPTIO.

Arbor mediocris. *Folia* binata vel ternata, biuncialia, canaliculata, margine nervoque scabra, apice subcarinata. *Vaginæ* breves, strictæ, minùs corrugatæ. *Amenta* nondum vidi. *Strobili* solitarii, recurvato-penduli, angustè ovati, muricati, spinis subincurvatis, squamis medio dilatatis.

I HAVE never seen more than two trees of this species in England; one at Pain's Hill, where I procured specimens for the engraving; the other at Kew.

The native situation of *P. variabilis* is the sea-shore of North America, or at no great distance from it, in a sandy, but mixed kind of soil. In New York under the forty-first degree of north latitude, its height is seldom above forty feet, and the shaft or trunk, not more than fifteen or twenty, parting then into branches pretty distant from one another. The *bark* is brownish and deeply cracked. The *wood* has a spunginess and lightness which deprives it of durability, and renders it useless in building, or indeed for any purposes of a similar kind, but it is tolerably full of resin, so that the Americans employ it for its tar and pitch. The *leaves* are two inches long, and pointed; in colour, dark green. The *flowers* appear at the beginning of May, and the seed ripens in November. The *cone* is about three inches long, and two thick at the base, it is rather bent at the top. The scales have a yellowish-brown tinge, and there are thorny points of a strong woody texture projecting from them. The seeds are smaller than in *P. sylvestris*.

EXPLANATION OF TAB. 15.

a, a.   Leaves with their sheaths.
B.   Point of a leaf.
c.   Ripe Cone.
d, d.   Scales of the same.
e.   Seed.

Pinus variabilis.

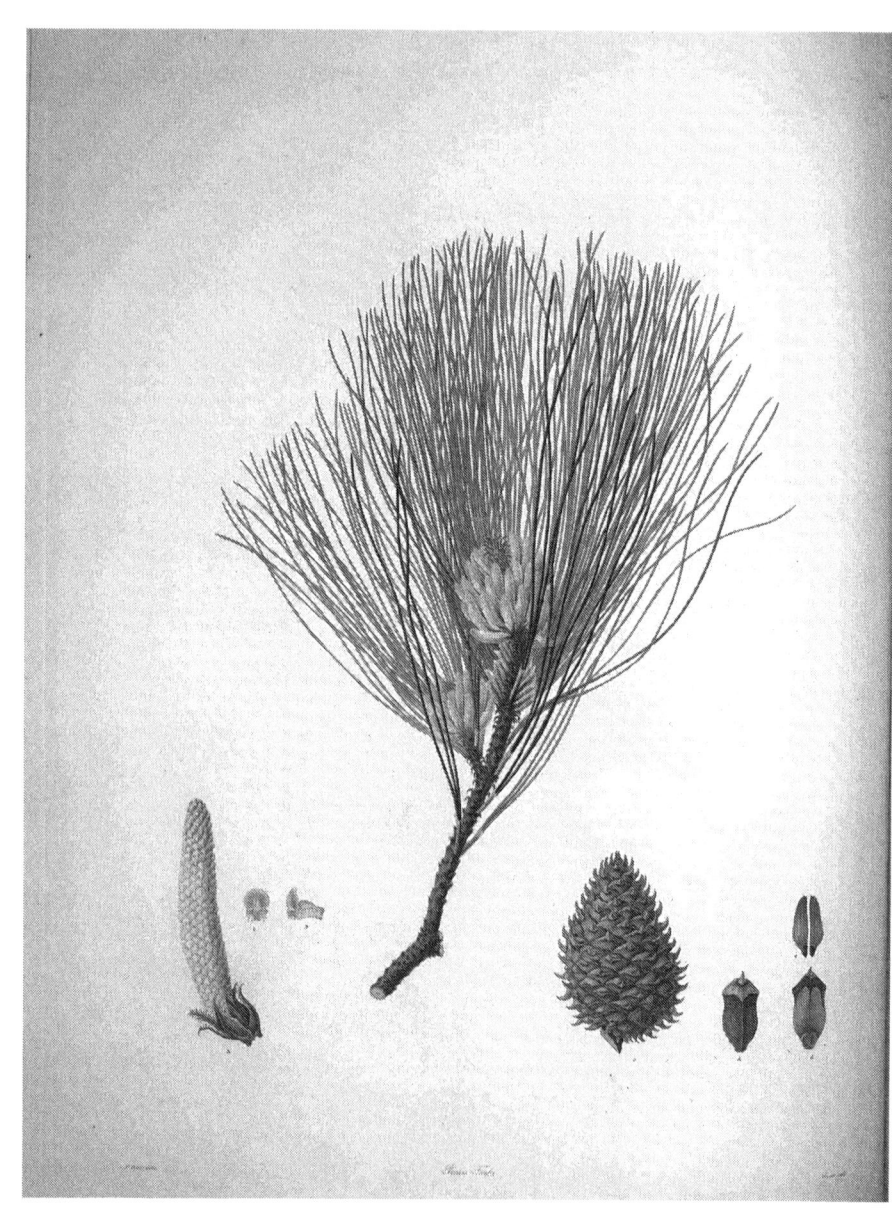

Tab XII.

Pinus Taeda

TAB. 16 & 17.

12. PINUS TÆDA.

FRANKINCENSE PINE.

PINUS TÆDA, foliis ternis elongatis, strobilis deflexis; spinis inflexis, vaginâ foliorum elongatâ.

P. *Tæda*, foliis trinis, conis oblongo-conicis, folio brevioribus, aggregatis, squamis echinatis. *Soland.*
MSS. *Ait. Kew. v.* 3. 368. *Linn. Sp. Pl.* 1419. *Syst. ed. Reich. v.* 4. 178.
P. *Tæda*, foliis ternis. *Gron. Virg.* 2. 152. *Wangenh. Beyt.* 41.
P. foliis longioribus tenuioribus ternis, conis maximis laxis. *Mill. Dict. n.* 11. *Evel. Sylv. ed.*
Hunter. 264.
P. foliis longissimis ex unâ thecâ ternis. *Colden. Novebor.* 230. *in Act. Soc. Reg. Sc. Ups.* 1743.
P. *Tæda*, foliis ternis, conis pyramidatis; squamis oblongis obtusis reflexis. *Du Roi. Harbk. ed. Pott.*
v. 2. 63.

Habitat in Americâ Septentrionali.
Floret Maio.

DESCRIPTIO.

Arbor humilis, ramosissima. *Folia* omnia ternata, spithamæa, angustissima, suprà nervo elevato, minùs canaliculata, subtùs planiuscula, margine scabra. *Vaginæ* unciales, parùm rugosæ, apice dilatatæ et laceræ. *Amenta mascula* cylindracea, densa; antherarum crista orbiculata, repanda, antheris paulò latior; *fæminea* subsessilia, ovata. *Strobili* sæpiùs bini, recurvato-patentes, ovati, acutiusculi, muricati, spinis incurvato-patentibus, acutissimè pungentibus.

PLAINS consisting of dry sand, and sea coasts, in North America, are abundantly stocked with this species of Pine. As the soil is too meagre to afford much nourishment, the trees growing in it (as we are informed by Wangenheim) remain low, full of branches, and attain but moderate strength. Their wood is of short duration, being apt to become worm-eaten, and rotten. If they grow, however, in moist and low places, they rise to a considerable height and strength, and occasionally afford timber for ship-building. The *bark* is greyish, rough, and cracked on old trees; the *leaves* generally measure from five to six inches in length, but often more. They are pointed, flat on the upper surface, furrowed on the under, pliable, and of a light green colour; there are always three in one sheath.

This is certainly a very distinct species from *P. rigida* (as Miller has made it), differing essentially in the loose texture of its cones, their slight attachment to the branches, the incurvature of the spines of the scales, and the length of the sheaths or *Vaginæ* of the leaves, which are longer than in *P. rigida*. The cones of the latter are of a much harder texture, and will remain on the

N

tree for several years, requiring some degree of force to detach them from it, whereas the cones of *P. Tæda* appear to fall off the tree soon after they are ripe. The *flowers* appear in Pensylvania, under the 40th degree of north latitude, towards the end of August. The *cones*, which require almost two years to arrive at maturity, are pyramidal, and from two to four inches long. The *scales* shoot into a woody, inflected point, and contain two *kernels* which are less than those of *P. sylvestris*, and ripen at the end of November, but the cones open and drop their seed only in warm weather. I could never find any male flowers on either of the two trees at Sion House, though they are so flourishing. Perhaps this circumstance is to be attributed to the dryness of the situation, and the lightness of the soil.

In regard to climate, our winters would seem to suit this species extremely well, and if it could be made to thrive on some of our heaths, the cultivation would be advantageous, if it were only for the tar, pitch, and turpentine. But to plant it in good soil would be unprofitable, because other pines planted under such circumstances, are far preferable on account of their greater durability.

EXPLANATION OF TAB. 16 & 17.

Tab. 16 is taken from a specimen brought from America by Mr. John Fraser.

    A. Male Catkin magnified.

    B, B. Antheræ.

    c. Ripe Cone.

    d, d. Scales of the same.

    e. Seeds.

Tab. 17, from a tree in the garden of the Duke of Northumberland at Sion House.

    a. Unripe Cones.

    b. Ripe Cones in their proper position.

    c. Scale.

    d. Seed.

See Cone of *P. Tæda alopecuroidea, Hort. Kew.* at *Tab.* 19. *fig.* 5.

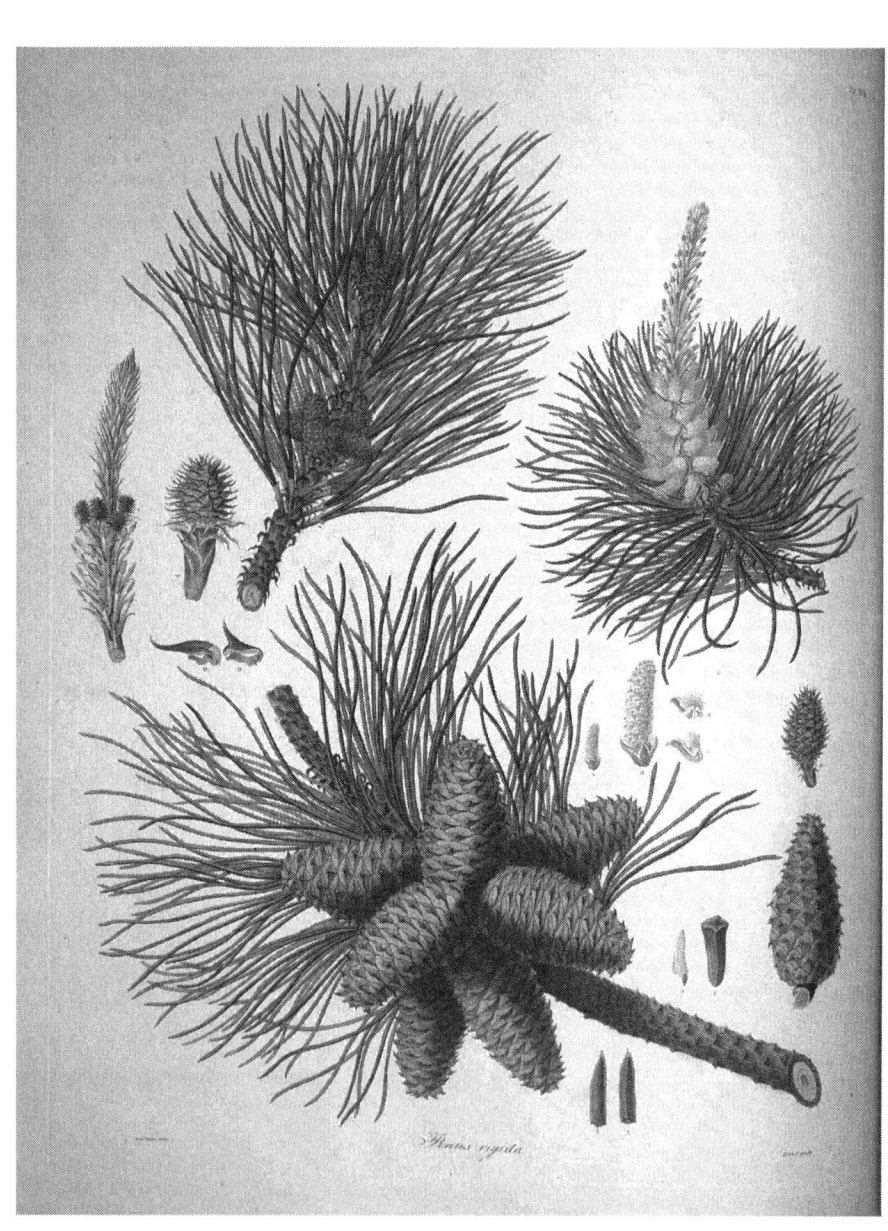

Pinus rigida

Tab. XIX.

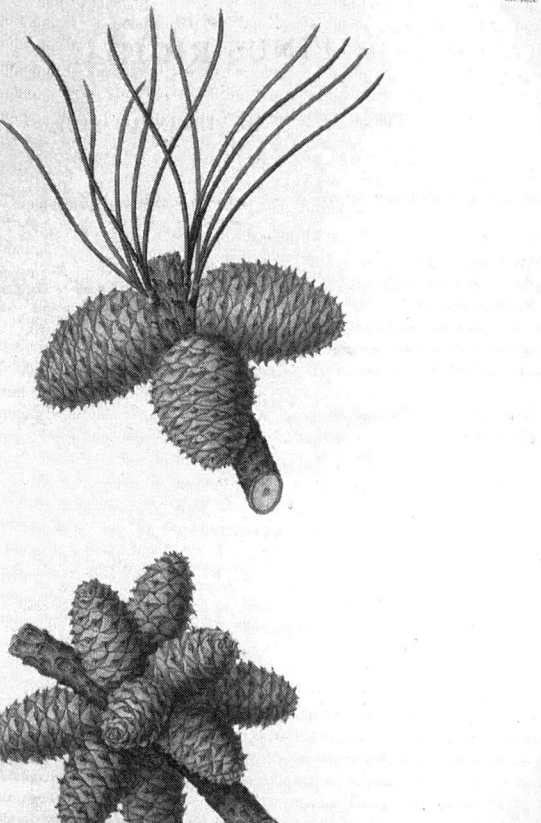

Pinus rigida

TAB. 18 & 19.

13. PINUS RIGIDA.

THREE-LEAVED VIRGINIAN PINE.

Pinus rigida, foliis ternis, strobilis ovatis confertis; squamarum spinis reflexis, vaginâ foliorum abbreviatâ.

P. Tœda rigida *g*. *Ait. Kew. v. 3. 368. Wilden. Berl. Baumz.* 210.

P. *rigida*, foliis ternis, conis pyramidatis, confertis; squamis oblongis acutis. *Du Roi. Harbk. v.* 2. 60. *Wangenh. Beyt.* 41.

P. *rigida*, foliis ternis, conis longioribus, squamis rigidioribus. *Mill. Dict. n.* 10.

P. canadensis trifolia, conis aculeatis. *Duhamel. Arb. v.* 2. 126. *n.* 16.

P. rigida. *Marshall. Arb. Amer.* 101.

Habitat in Americâ Septentrionali.
Floret Maio.

DESCRIPTIO.

Arbor excelsa. *Folia* breviora, latiora, & rigidiora quam in præcedente, at nervo margineque similia. *Vaginæ* vix semi-unciales, imbricato-rugosæ. *Amenta mascula* crassiuscula, purpureo-lutea, densa; antherarum crista reniformis, repanda, antheris latior: *fœminea* ovato-subrotunda. *Strobili* aggregati, undique patentes, ovati, magnitudine varii, muricati, spinis reflexis, acutis.

This Pine grows in Virginia, Maryland, and Pensylvania more abundantly than in districts further northward. Mr. Menzies found it in California, and I have been favoured by that gentleman with a large, full-grown cone gathered on the coast of that country, during his circumnavigation with Captain Vancouver, who brought home an immense collection of plants and drawings collected in that voyage. In growth and strength *P. rigida* is equal to *P. sylvestris*, but the wood is more spungy, and used for ships, and other buildings, only for want of better. It does not usually inhabit mountainous places, but dry sandy plains. The *leaves* distinguish the species sufficiently from any other, being from two to three inches long, pointed, smooth on the under surface, and furrowed above; there are very fine serratures on the edges, and the whole texture is strong. The sheaths of the leaves are much shorter than in *P. Tœda*. In Pensylvania, says Wangenheim, the *flowers* appear at the beginning of May, and ripe *seed* is procured in October. The *cones* are of a yellowish-brown colour, tapering, and a little curved toward the top. Their length is between three and four inches. Every *scale* has a woody point, which is reflexed. There is a variety of this species with smaller cones, some trees of which are growing at Pain's Hill. See Pl. XIX. f. 1, 2, 3, 4.

a

EXPLANATION OF TAB. 18 & 19.

TAB. 18, a, A. Male Catkin.
  B, B. Antheræ.
  c, C. Female Catkin.
  D, D. Scales.
  e. Unripe Cone.
  f. Ripe Cone.
  g. Scale.
  h. Seed.
  I, I. Points of Leaves.

TAB. 19. fig. 1—4. Cones of the small-fruited variety of *P. rigida* from Pain's Hill.
5. Cone of *P. Tæda alopecuroidea, Hort. Kew,* from a specimen preserved in spirits in the Banksian collection.

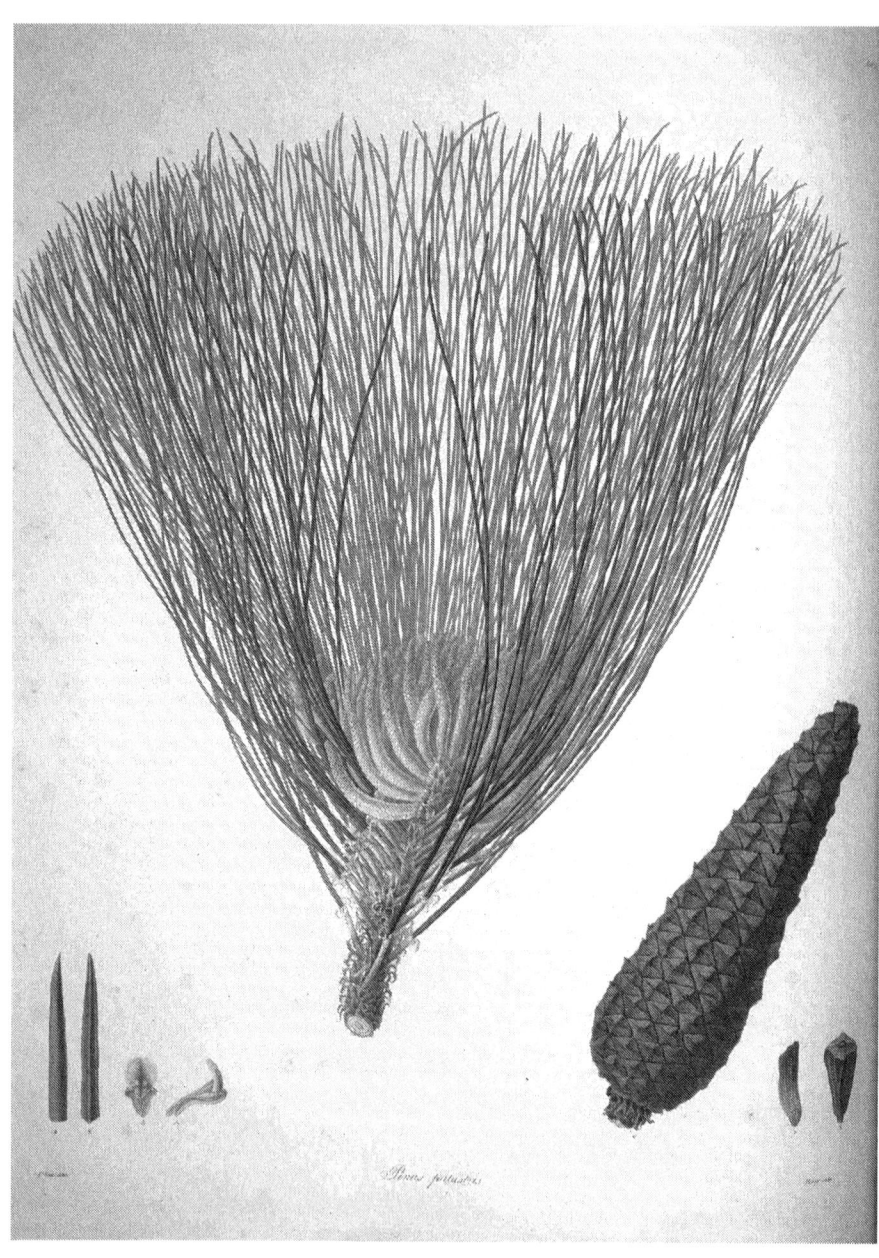

Pinus palustris

TAB. 20.

14. PINUS PALUSTRIS.

SWAMP PINE.

PINUS PALUSTRIS, foliis ternis longissimis, strobilis sub-cylindraceis muricatis, stipulis pinnatifidis ramentaceis persistentibus.

P. palustris. *Soland. MSS. Ait. Kew. v. 3. 308. Willden. Berl. Baums. 211.*

P. *palustris*, foliis ternis, conis oblongo-pyramidatis: squamis oblongis obtusis. *Du Roi. Harbk. ed. Pott. v. 2. 66.*

P. *palustris*, foliis ternis longissimis. *Mill. Dict. n. 14. Wangenh. Beyt. 73.*

P. americana palustris, trifolia, foliis longissimis. *Hort. Angl. 88. Duhamel. Arb. v. 2. 126. n. 18.*

P. palustris. *Marsh. Arb. Amer. 100.*

Habitat in Americæ Septentrionalis paludosis.

DESCRIPTIO.

Arbor mediocris, gracilis. *Folia* longissima, ferè pedalia, undique patentia, formosa, subtùs convexa, ecarinata, suprà nervo prominulo à basi ad apicem notata, margine nervoque scabra. *Vaginæ* unciales, corrugatæ, laceræ. *Stipulæ* elegantèr pinnatifidæ, persistentes. *Amenta mascula* cylindracea, elongata, purpurascentia; antherarum crista rotundata, convexa, subdenticulata, antheris angustior: *fœminea* nondùm vidi. *Strobili* spithamæi, subcylindracei, recti, tuberculoso-muricati, spinis brevibus, incurvis, obsoletis.

I AM indebted to Mr. Fairbairn of Chelsea Gardens for the cone, from which that in the plate was drawn, and for the branch of Male Flowers to that indefatigable collector Mr. John Fraser, who in three several voyages to America brought back each time a large collection of plants; and by whose means many new species adorn our gardens. P. *palustris* grows only in the warm and moderate climates of North America. Wangenheim found it as far northward as forty degrees latitude in Pensylvania; but there, he remarks, it is generally solitary and the offspring of cultivation. In Virginia and Carolina it grows in greater numbers. Dry, elevated land does not seem to suit it, but low marshy spots sufficiently sheltered, says Wangenheim. Its height is between forty and fifty feet, and diameter of the trunk nearly two; in proportion therefore to other species, this tree is inconsiderable. The *bark* is grey and much cracked upon old trees. The *wood* is of a reddish white colour, soft, light, and very sparingly impregnated with resin; it soon decays, and burns badly. It is so little esteemed, that as long as any other kind of wood is to be had, not the least use is made of it. When swamps are dried up and prepared for cultivation, all the trees of this kind growing in them are consumed on the spot.

P

The *leaves* stand closest towards the terminations of the branches, they are from eight to twelve inches long, slender, and of a light green.

The habit of this tree being very singular and curious, and so different from any of its congeners, it forms an interesting appearance, and ought to be planted, in every *arboretum*; but, as it is one of the most tender of the *genus*, it should be well sheltered from the cold, until it is become of a large size; owing to this neglect, almost all the trees which have been planted in this country have been lost, and I know of only two of any size remaining, one at the Royal Gardens at Kew, the other at the Earl of Coventry's.

EXPLANATION OF TAB. 20.

A, A. Anthers.
B, B. Points of leaves.
c. Ripe Cone.
d. Scale.
e. Seed.

Pinus longifolia.

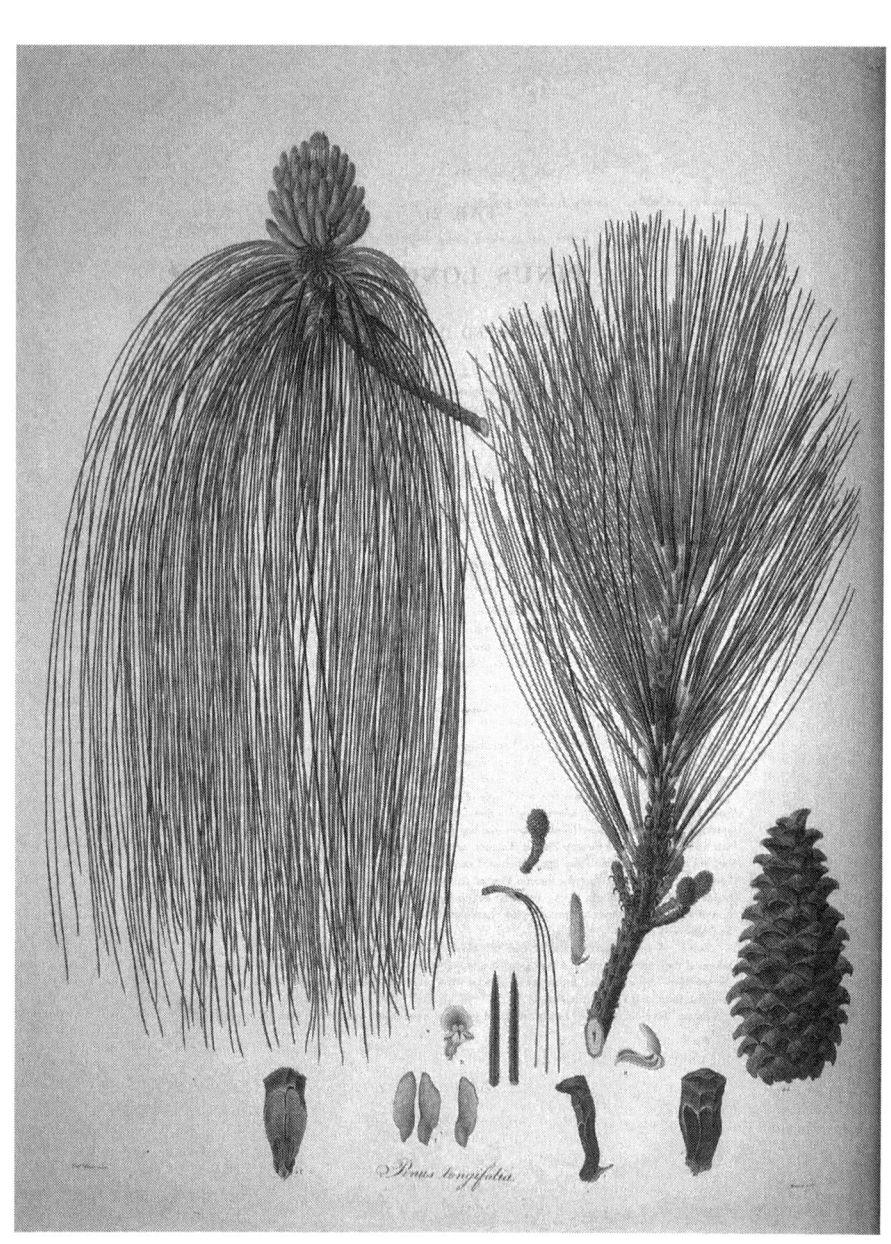

Pinus longifolia.

TAB. 21.

15. PINUS LONGIFOLIA.

LONG-LEAVED INDIAN PINE.

PINUS LONGIFOLIA, foliis ternis tenuissimis longissimis pendulis, vaginâ elongatâ, stipulis integerrimis deciduis, antherarum cristâ convexâ integriusculâ.

P. longifolia. *Roxb. MSS.* 1180.

Habitat in montibus Napaulensibus, Indiæ Orientalis.

DESCRIPTIO.

Arbor vasta, excelsa. *Folia* demùm pendula, ultrà pedalia, gracillima, subtùs striata, convexa, suprà canaliculata, nervo tenui, prominulo, margine undique serrulato-scabra. *Vaginæ* semiunciales, vel paulò longiores, læves, apice laceræ. *Stipulæ* breves, integerrimæ, recurvæ, deciduæ. *Amenta mascula* ovato-cylindracea, duplò quam in præcedente breviora; antherarum crista priori similis sed latior: *fœminea* globosa, pedunculata, erecta. *Strobili* ovati, parùm incurvi, tuberculosi, nec muricato-spinosi, haud biunciales.

For the following account of this tree, I am obliged to a manuscript communication of Dr. Roxburgh's. " *Leaves* threefold, filiform, very long and pendulous, with margins a little scabrous. *Cones* ovate, considerably shorter than the leaves, *scales* thereof smooth, *anthers* crowned.

In gardens about Calcutta a few small trees of this species are found, all from Napaul, or reared from seed from that country, where they are found on the stupendous mountains, there growing to an immense size, and there they blossom about the beginning of the hot season. *P. Tæda*, with its varieties, and *P. palustris*, or Swamp Pine of America, are the only other species with threefold leaves; but as I am not in possession of any figure thereof, I cannot take upon me to say, this is not one of them; however it is not likely that the Swamp Pine of America, should be found an Alpine plant in Asia; besides, their cones differ in shape; the great length and disposition of the leaves, as well as the structure of the scales of the cone, preclude the chance of its being *P. palustris*, or *P. Tæda*, or any one of its (supposed) varieties.

Trunk. I have observed above, that the trees about Calcutta are small, but in Napaul I am informed they grow straight to a very great height, upwards of an hundred feet; the bark is scabrous, the branches verticelled, and rather few in number than otherwise, so that here the head is thin, of a roundish form, and yields little shade.

Leaves threefold, disposed in approximated spiral rows round the ends of the branchlets, perfectly filiform, margin somewhat hispid when the finger is pulled backward, generally pendulous, and from nine to eighteen inches or more in length.

Stipules or Sheaths round the base of the leaves numerous and chaffy.

MALE FLOWERS.

Antheral racemes numerous at the extremities of the branchlets: from their centre issues the shoot of the same season.

Bracts solitary, one to each raceme.

Calix of the raceme, scales numerous, round, its base chaffy, brown.

Stamens very numerous. *Filaments* scarce any. *Anthers* clavate, opening on each side and crowned with a large roundish scale inflected over the next above."

EXPLANATION OF TAB. 21.

a,   Male Catkin.
B, B. Antheræ.
c.   Young Cone.
d, d. Ripe Cone.
e.   Scale of the Cone.
f.   Seed.
g.   Sheath of the Leaves.
H, H. Points of the Leaves.

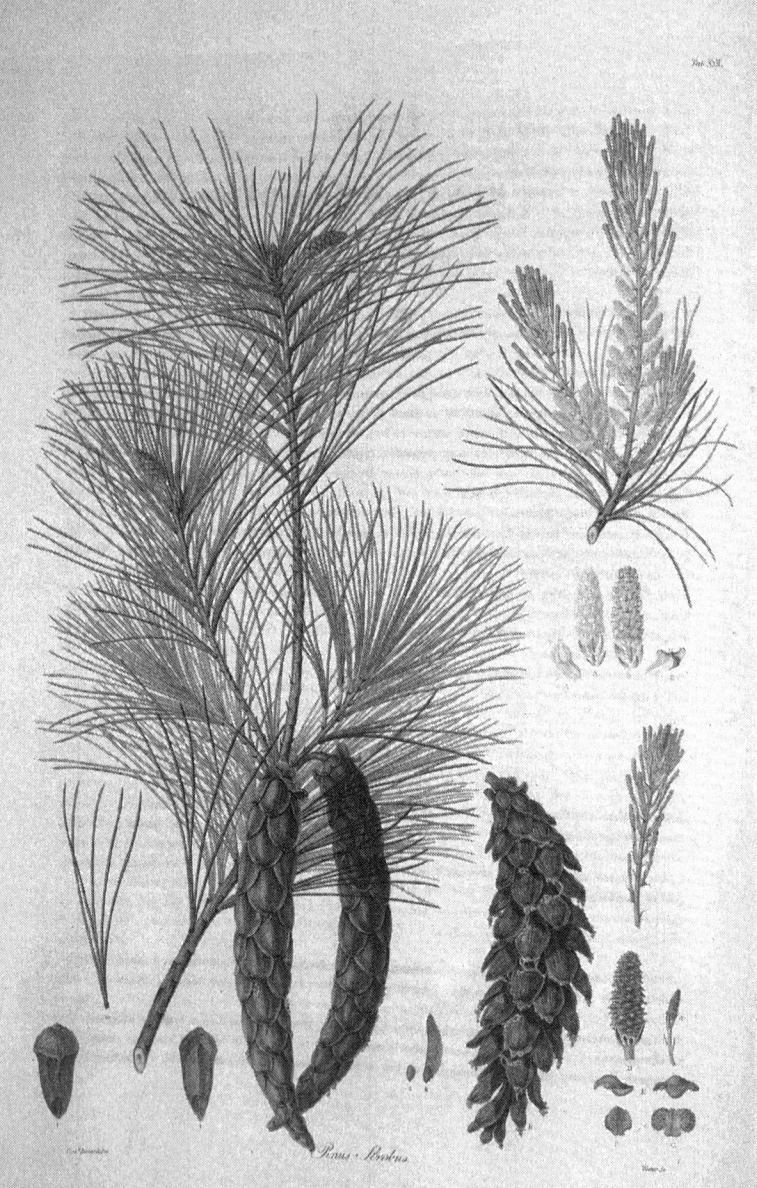

Tab. XII.

Pinus Strobus

TAB. 22.

16. PINUS STROBUS.

WEYMOUTH PINE.

PINUS STROBUS, foliis quinis, strobilis folio longioribus cylindraceis lævigatis, antherarum cristâ geminâ subulatâ minimâ.

P. *Strobus*, foliis quinis, conis cylindraceis, folio longioribus, laxis. *Soland. MSS. Ait. Kew. v.* 3. 369.

P. *Strobus*, foliis quinis margine scabris, cortice lævi. *Linn. Sp. Pl.* 1419. *Syst. ed. Reich. v.* 4. 174. *Mill. Dict. n.* 13. *Evel. Sylv. ed. Hunter.* 263. *Wangenh. Beyt.* 1. *t.* 1. *f.* 1. *Vitm. Sp. Pl. v.* 5. 345.

P. *Strobus*, foliis quinis, conis oblongis pendulis, squamis ovalibus planis laxis. *Du Roi. Harbk. ed. Pott. v.* 2. 78.

P. Strobus. *Marsh. Arb. Amer.* 101.

P. foliis quinis cortice glabro. *Gron. Virg.* 2. 152.

P. foliis longissimis, ex unâ thecâ quinis. *Colden. Novebor.* 229. *in Act. Soc. Reg. Sc. Ups.* 1743.

P. canadensis quinquefolia, floribus albis, conis oblongis et pendulis, squamis abieti ferè similis. *Duham. Arb. v.* 2. 127. *n.* 19.

P. virginiana, conis longis, non ut in vulgari, echinatis. *Pluk. Alm.* 297.

Larix canadensis, longissimo folio. *Tournef. Inst.* 586.

Die Tannen fichte. *Linn. Pfl. Syst. v.* 2. 355.

Habitat in Americâ Septentrionali.
Floret Maio.

———————

DESCRIPTIO.

Arbor excelsa, recta, cortice lævi, cinereo, ætate resinoso. *Folia* quinata, patula, triuncialia, gracilia, subtùs carinato-triquetra, margine scabra. *Vaginæ, Stipulæque* nullæ. *Amenta mascula* elliptica, brevia, pallidè purpurascentia; antherarum crista omnium minima, é setis duabus erectis, brevissimis: *fœminea* ovato-cylindracea, erecta, breviùs pedunculata. *Strobili* penduli, cylindracei, incurvati, læves atque glaberrimi.

———————

" *Cotyledones* 6 ad 10. *Folia* quina, molliora, triquetra, scabra, angulis inconspicuè serratis. *Strobili* spithamæi, apice squamarum crassiore; hæc conjungit Pinum & Abietem Auctorum." *Reich.*

THE specimen here figured was taken from a fine tree growing in the Royal Garden at Kew. Of all the species of *Pinus* hitherto known, the Weymouth Pine grows the highest, the straightest, and strongest, and may justly be considered as the chief of this numerous tribe. It inhabits in most perfection the extensive territory comprehended between the forty-second and forty-fifth degree of north latitude. Within this space lie large portions of the provinces of New York, New England, Nova Scotia, and Canada. The principal woods are on the shores of Fundy Bay and of Casco Bay (in Nova Scotia) on the eastern side of Massachuset Bay; on the shores of the rivers Piskatoqua and Merrimach (in New Hampshire) and of the Connecticut, and Mohawk; and from the extreme northern side of the river St. Lawrence towards Montreal and the shores of the Lake Champlain. In the more southerly parts of North America, this species appears but sparingly (as Wangenheim informs us) and never in continued forests; a proof that a cold, rough climate suits it best. The soil in which this Pine is found, is said to be of the best kind, being a clay mixed with sand and other earths; it is light and moist, preserving these qualities to the depth of some feet.

The vallies, the crevices of the mountains, and banks of rivers are the conservatories, as it were, to which the rains and melted snows in the spring carry down the fattest parts of the soil of the higher lands. In these spots, which are sometimes pretty elevated, the natural plantations of *P. Strobus* are seen growing to a height and thickness, not exceeded by any other of the tribe; indeed, few come near to it in these particulars. It is certain that among the full-grown trees, on the best ground, there are some two hundred feet in height, and four or five in diameter at the lower end of the trunk. Wangenheim tells us that he was convinced of the truth of this statement when he was in the dock-yards of Plymouth. "We saw," says he, "two masts for seventy-four gun ships which measured in the whole piece one hundred and eight feet in length, and a roller that was every where three feet in diameter. Such a tree must have been two hundred feet long, and five feet or more in diameter." *(Beyt. p. 2)*. The growth of this tree, as we are informed by the same intelligent author, is very uniform in its native forests wherever it is surrounded with others. It naturally prunes itself, the branches falling off of their own accord. When the young tree stands free, and exposed on open spots, the branches are very extensive, and the planks cut from such have no knots. But when several are growing close together they attain their full size in sixty years, whereas the same height in *P. Picea* usually requires one hundred years. Under these circumstances, in advanced years *P. Strobus* has a very small top, in proportion to its height and thickness, composed of long twigs, which do not break under the pressure of the heaviest snow, a pressure that would otherwise greatly impede the growth of the tree. The *bark*, at first, is pretty smooth and of a dark grey colour, but in old trees it becomes somewhat brown and abundantly impregnated with a whitish resin, which has a very agreeable odour. The *wood* is of a yellowish white colour, of a tolerable hardness, very fine, almost resembling the white cedar, and works straight, smooth, and shining. It contains many volatile resinous particles, which contribute greatly to its preservation. The *leaves* are almost three inches long, and grow five in a sheath. They are of a bright green colour, triangular, and very finely serrated. The *flowers* appear at the end of April. The *cones* are from six to eight inches long, and nearly one inch in diameter. Every one of them has a short fruit-stalk, and two or three generally shoot round the same branch. Before the cones open, the *scales* lie loosely upon one another. The latter are round, smooth, and, when ripe, of a brownish copper colour; at a distance they assume a whitish cast occasioned by the sun melting a kind of turpentine which oozes from the unripe scales, and makes them clammy. The *seeds* have uniform wings, and are glued, as it were, to the scales by the resinous exudation. When the seed flies out, the wings are generally broken by the wind, so that it is not usually carried far from the tree. It ripens towards the end of August, and, if there happen to be hot days about the middle of September, it will be shaken out.

When there is an intention to rear considerable plantations of *P. Strobus*, good seed should be carefully chosen, and a soil prepared neither too rich nor heavy, and mixed with sand. As to transplanting, every able nurseryman will allow that it impedes the natural growth; he will have remarked also that plants set at a distance from one another grow more to twigs, prune themselves later, and therefore in an equal number of years, rise to less height than if they had been planted close together. Experience shews that the English climate is suited to the growth of this tree; the situation however is a matter of much importance; cold and mountainous spots are certainly the most desirable, and yet they should be such as are protected against violent tempests; for instance, crevices and recesses of mountains, for when too much exposed it is very liable to be injured by the cold winds. The soil in which the seed is to be sown in the spring, should be turned up, and, if possible, the clods broken shortly before the preceding winter. The end of March, or the beginning of April, seems to be the proper period for sowing, and the seed should not be covered over. If sown in the Autumn, it will be necessary to turn up the ground immediately before. With respect to cutting, this should take place when the trees have acquired their full height and strength, and the spots which are cleared may be sown with seed of home produce. Besides furnishing timber for all sorts of masts and yards, and for a considerable part of the hull of a ship, as far as planking is required, this pine is of the greatest use to the common carpenter, who can turn almost every part of it to account. The North American has discovered its value, though hitherto but little skilled in forest botany, and studiously preserves the young trees from the depredations of cattle. The wood lasts as long above ground as that of any known species of *Pinus*, but in building under ground, for door thresholds, and for the hulls of ships, it should be used only in cases of necessity, as its duration in such situations is rather short, and there are other woods better adapted to such purposes. It yields a very fine resin from which good turpentine may be prepared. The earliest propagation of *P. Strobus* in this country was at Lord Weymouth's (from whom it had its name) in Wiltshire, and at Sir Wyndham Knatchbull's in Kent. Most of the seeds, afterwards sown, were procured from these places, so that our island may be said to have been stocked from them. Although none of the Pines (except the Larch tribe) are deciduous, yet the position of the leaves becomes very different in Winter, from what they are in Summer; in the latter they are much more divaricated, in the former they become nearly parallel to the stem. (*Folia adpressa, et folia divaricata.*) In no species of Pine is it more exemplified than in this.

EXPLANATION OF TAB. 22.

A, A. Unripe male Catkin with an unopened Anthera.
B, B. Ripe male Catkin, and Anthera which has shed its pollen.
C, C. Crest.
d, D. Female Catkin.
E. Scales of the same.
F. Upper Scale separate.
G. Under Scale.
h. Ripe Cone.
i, i. Scales of the same.
k. Seed with its wing.
l. Seed stripped of its wing.
m. Leaves.

8

TAB. 23 & 24.

17. PINUS CEMBRA.

SIBERIAN STONE PINE.

Aphernouilli of the Swiss.

PINUS CEMBRA, foliis quinis, strobilis ovatis, seminum alis obliteratis, antherarum cristâ reniformi crenatâ.

P. *Cembra*, foliis quinis, conis ovatis obtusis, squamis adpressis, nucibus duris. *Soland. MSS. Ait. Kew. v. 3. 369. Willden. Berl. Baumz. 212.*

P. *Cembra*, foliis quinis lævibus. *Linn. Sp. Pl. 1419. Syst. ed. Reich. v. 4. 173. Fl. Scan. 32. Mill. Dict. n. 6. Scop. Ann. 2. 65. Evel. Sylv. ed. Hunter. 265. Pall. Fl. Ross. v. 1. 3. t. 2. Allion. Fl. Ped. v. 2. 179. Vitm. Sp. Pl. v. 5. 344. Villars. Dauph. v. 3. 806.*

P. foliis quinis, cono erecto, nuce eduli. *Gmel. Sib. v. 1. 179. t. 39. Duhamel. Arb. v. 2. 127. n. 20. t. 32.*

P. foliis quinis triquetris. *Hall. Helv. n. 1659.*

P. foliis quinis, conis ovatis erectis, squamis ovalibus concavis, nucibus cuneiformibus, alâ membranaceâ destitutis. *Du Roi. Harbk. ed. Pott. v. 2. 69.*

P. sativa, cortice fisso, foliis setosis, subrigidis, ab unâ vaginâ quinis. *Amm. Ruth. 178.*

P. sylvestris montana tertia. *Bauh. Pin. 491.*

P. sylvestris Cembro. *Cam. Epit. 42.*

Larix sempervirens, foliis quinis nucleis edulibus. *Breyn. in Act. Nat. Cur. Cent. 7. 8. Obs. 2. t. 1. f. 3, 4, 5.*
Die Cedernfichte. *Linn. Pfl. Syst. v. 2. 353.*

Habitat in Alpibus Sibiriæ, Tartariæ, Helvetiæ, Vallesiæ, Baldi, Allobrogum, Tirolensium, Tridentinorum.
Floret Maio.

DESCRIPTIO.

Arbor erecta, robusta, tardè crescens. *Folia* sæpiùs quinata, quandoque quaterna vel sena, patentia, plerumque triuncialia vel paulò longiora, subtùs planiuscula, suprà nervo maximè elevato notata, margine scabriuscula. *Vaginæ* nullæ. *Stipulæ* lanceolatæ, acuminatæ, persistentes. *Amenta mascula* elliptica, pulcherrimè violacea; antherarum crista reniformis, dentato-erosa, antherâ angustior: *fœminea* elliptico-globosa, erecta, subsessilia. *Strobili* ovati, cernui, læves; juniores violacei cum rore glauco.

P. *Cembra*, Duhamel informs us, flourishes in the coldest parts of France, where it is called *Alviez*; but it is most abundant on the Swiss Alps, in spots covered with snow, and where no other vegetation flourishes. In Siberia, it seems to be most luxuriant in similar situations, but even marshy spots are not unsuitable to it, as we find from Gmelin, who pronounces this tree, "*frigoris patientissima, et locorum palustrium amantissima.*" From these facts we may infer that it may be planted on our bleak, and mountainous lands, and if these should be situated in the vicinity of the sea, that circumstance may

Tab XXX

3 Pinus Cembra

35

not be detrimental to its growth. When young, and in warm weather it will bear being transplanted, see Hart's Essays on Husbandry.

The timber of *P. Cembra* is large, and has a finer grain than common deal; its smell is remarkably pleasant. The *bark* of the trunk is of a whitish cast. The *leaves* are of a lighter green than most of the other species, and they closely ornament the branches all round. They are from three to four and a half inches long; the number, that springs from one sheath, is five. The *flowers* have a more beautiful appearance than in any other species, being of a bright purple colour, as are also the unripe, full-grown cones, which have a bloom upon them resembling that of a ripe Orleans plumb. The *cones* are usually almost two inches in diameter, their length is in general not more than three, and the scales are of an oval form, often reflexed at the margins. The *nuts* are triangular, and easily cracked, especially when ripe. The *kernels* are about the size of a common Pea, and have the whiteness and softness (when stripped of a brownish rind) of a blanched Almond. They have an agreeable oily taste, and often form part of a Swiss as well as of a Siberian desert; in the latter country, during a favourable season, such quantities are produced that the poorest peasants may provide themselves with many pounds at a very trifling expence.

In the plantations of Jeremiah Dixon, Esq. near Leeds, may be seen several Pines of this species, which, in that neighbourhood, is generally denominated the *Gleddow Pine*, from the place where it is cultivated. On Lord Clive's estate also, in Shropshire, there are very flourishing plantations of it: the seeds of these last were brought from Switzerland by Mr. Hyams, who kept the Florida gardens some time ago, and who after having supplied a few of the nurserymen with plants, sold the remainder of his stock (amounting to more than two thousand) to the above-mentioned nobleman. They are become a great ornament to the vicinity of Walcot. *P. Cembra* is one of the handsomest trees of the whole *genus*, but of the slowest growth, as may be seen from those at Mill Hill, the two largest of which are seventy years of age, and the smallest about fifty, as I have been informed by the gardener, who lived there a considerable time. At a younger period, their growth is still slower, for they seldom attain the height of three feet until their age amounts to fourteen years or more. I have seen trees of this height bear Male Flowers. This species at present is scarce in the Nursery Gardens about London, and bears a high price; it is to be regretted that more of this very ornamental tree has not been introduced into this country.

EXPLANATION OF TAB. 23 & 24.

TAB. 23, representing the Male Flowers and Ripe Cones, was taken from specimens procured in the Royal Gardens of Kew.

a, A. Male Catkin.
B. Anthera discharging its Pollen.
c. Young Cone.
d. Ripe Cone.
e, e. Its Scales.
f, f. Seeds.
g. Seed opened to shew the kernel.
h. Kernel, with a separate view of its basis.
i. Longitudinal section of the kernel.
k, K. Embryo with its radicle and cotyledons.
l. Leaves.
M. Point of a leaf.

TAB. 24, shewing the full-grown but unripe Cones, was delineated from a specimen in the garden of the late Mr. Peter Collinson, at Mill Hill.

a. Female Catkin.
b. Fully formed, but unripe, Cone.

18. PINUS OCCIDENTALIS.

WEST INDIAN PINE.

PINUS OCCIDENTALIS, foliis quinis longissimis: margine scabris, strobilis oblongis; squamis apice trun-
catis. *Swartz. Prod.* 103. *Flora Occid. v.* 2. 1230.

P. foliis quinis ab eodem exortu. *Plum. Cat.* 17. *Plant. Amer.* 154. *tab.* 161.

Habitat in Montibus Hispaniolæ, Quartier du Pin.

DR. SWARTZ seems only to have seen trees of this species without male flowers or fruit, and could only procure a branch with leaves, and a cone very much mutilated; it therefore still remains to be better described by some future Botanist who may be more fortunate.

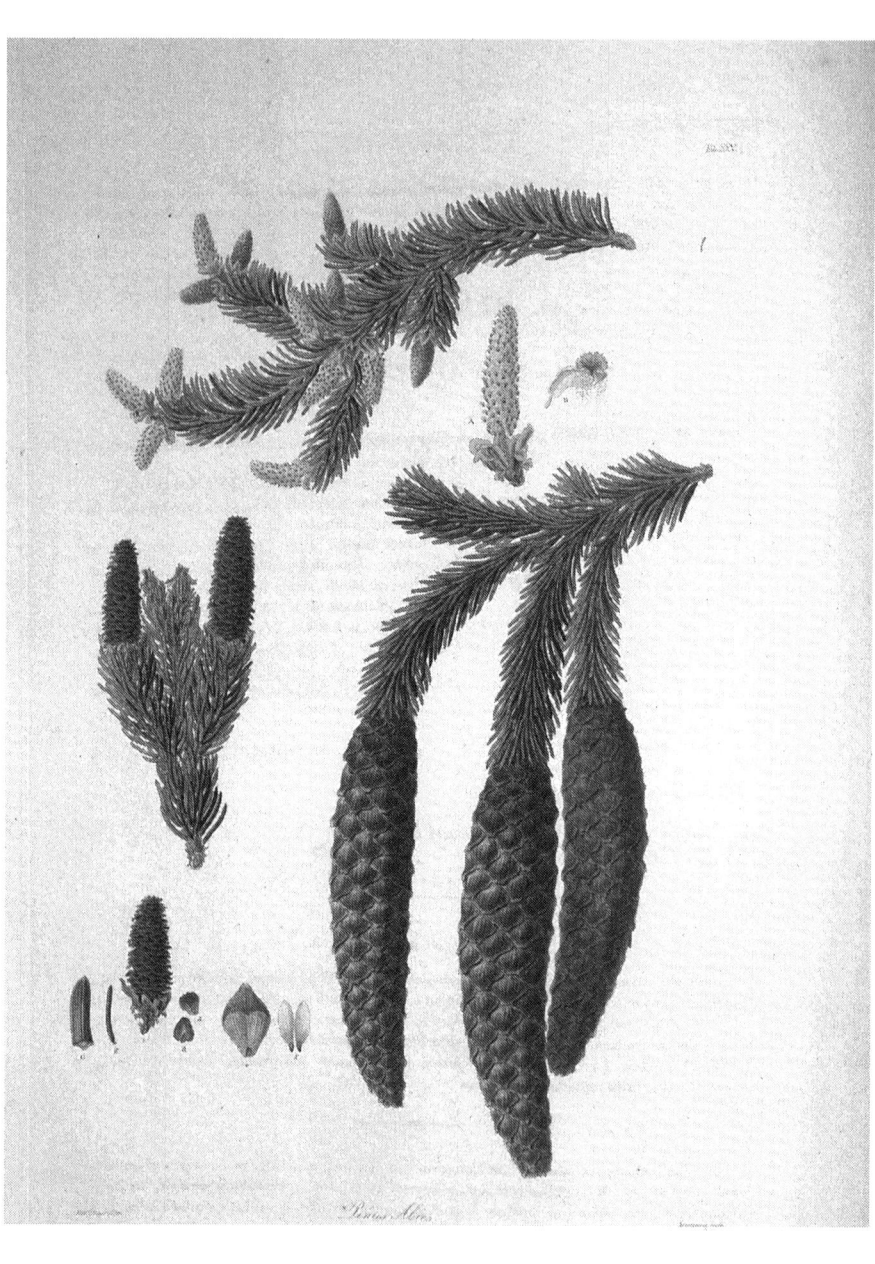

Pinus Abies

TAB. 25.

19. PINUS ABIES.

NORWAY SPRUCE FIR.

PINUS ABIES, foliis solitariis tetragonis, strobilis cylindraceis; squamis rhombeis complanatis margine repandis erosis.

P. *Abies*, foliis solitariis subtetragonis acutiusculis distichis, ramis infrà nudis, conis cylindraceis. *Ait. Kew. v.* 3. 371. *Willden. Berl. Baumz.* 221.

P. *Abies*, foliis solitariis subulatis mucronatis lævibus bifariam versis. *Linn. Sp. Pl.* 1421. *Syst. ed. Reich. v.* 4. 177. *Fl. Suec. n.* 875. *Lapp. n.* 347. *Mat. Med. n.* 473. *Huds. Angl.* 424. *Scop. Carn. n.* 1194. *Gunn. Norv. n.* 39. *Evel. Sylv. ed. Hunter.* 266. *Trew. in Nov. Act. Ac. N. Cur. v.* 3. *App.* 445. *t.* 14. *f.* 5. 10. *et t.* 16. *f.* 1. 10. *Mattusch. Sil. n.* 705. *Dörr. Nerz.* 263. *Blackw. t.* 198. *Fl. Dan. t.* 193. *Regn. Bot. Pall. Fl. Ross. v.* 1. 6. *t.* 1. *f.* G. *Allion. Fl. Ped. v.* 2. 180. *Woodv. Med. Bot.* 573. *t.* 208. *Vitm. Sp. Pl. v.* 5. 346. *Villars. Dauph. v.* 3. 810.

P. foliis solitariis, tetragonis, mucronatis. *Hall. Helv. n.* 1656.

P. picea, foliis solitariis subulatis bifariàm versis, conis oblongis pendulis, squamis ovalibus planis marginibus undulatis et laceris. *Du Roi. Harbk. ed. Pott. v.* 2. 156.

Abies picea. *Mill. Dict. n.* 2.

A. foliis solitariis apice acuminatis. *Linn. Hort. Cliff.* 449. *Fl. Suec. ed.* 1. 789. *Fl. Lapp. ed.* 1. *n.* 347. *Dalib. Paris.* 295. *Gmel. Sib. v.* 1. 175.

Die Bothtanne. *Linn. Pfl. Syst. v.* 2. 368.

Habitat in Europæ, Asiæ borealibus humidiusculis.
Floret Aprili.

DESCRIPTIO.

Arbor excelsa, recta, pyramidalis, ramis inferioribus deflexis. *Folia* solitaria, undique patula, uncialia, tetragona, obtusiuscula, lævia, nitida. *Stipulæ* nullæ. *Amenta mascula* erecta, breviùs pedunculata, ovato-cylindracea: antherarum crista reniformis, dentato-lacera, antherâ parùm angustior: *fœminea* cylindracea, erecta, solitaria, terminalia, sessilia; bracteolæ squamis interstinctæ, minimæ, mucronulatæ. *Strobili* penduli, 4 vel 5 unciales, cylindracei, læves, squamis latè ovalibus, seu rhombiformibus, complanatis, extùs repandis, apice erosis.

P. *Abies* is one of the loftiest of the European trees, growing sometimes to the height of one hundred and fifty feet. It is commonly straight and pyramidical. The *bark* is reddish and scaly, the *leaves* shoot very thickly, but not so regularly as in P. *picea*; they are slightly carinated on both sides, of a dusky green colour, shining on the upper surface, and often curved. In summer, after a long continuance of dry weather, I have seen most of them decay and fall off.[*] The *cones* are nearly cylindrical, of a purple colour, and sometimes green before they are ripe, always pendent. The *scales* assume an oval shape, and become somewhat ragged on the edges. The *seeds* are small, rather flattened, and oval, with two thin elliptical membranous wings.

The wood of *P. Abies* is extremely serviceable for a great variety of purposes, being very firm, straight, and regular in the grain, and capable of resisting moisture a long time: that which is grown in England is said to be more durable than what is imported, and to be particularly useful in making of ladders. From the resin, yielded by this tree, the Burgundy pitch is prepared. The insects commonly inhabiting *P. Abies*, are *Phalæna strobilina*, *Chermes Abietis*, and *Cimex abietinus*.

EXPLANATION OF TAB. 25.

Pain's Hill afforded me the specimen for the engraving.

A.   Male Catkin.

B.   Anthera.

c.   Female Catkin.

d, d. Scales of the same.

e.   Scale of the ripe Cone.

f.   Seeds.

g, G. Leaf.

[*] Many curious Pines were lost at Sion-House, some years ago, in consequence of standing in very dry and hot situations; among them was *P. maritima*.

Tab. XXVI

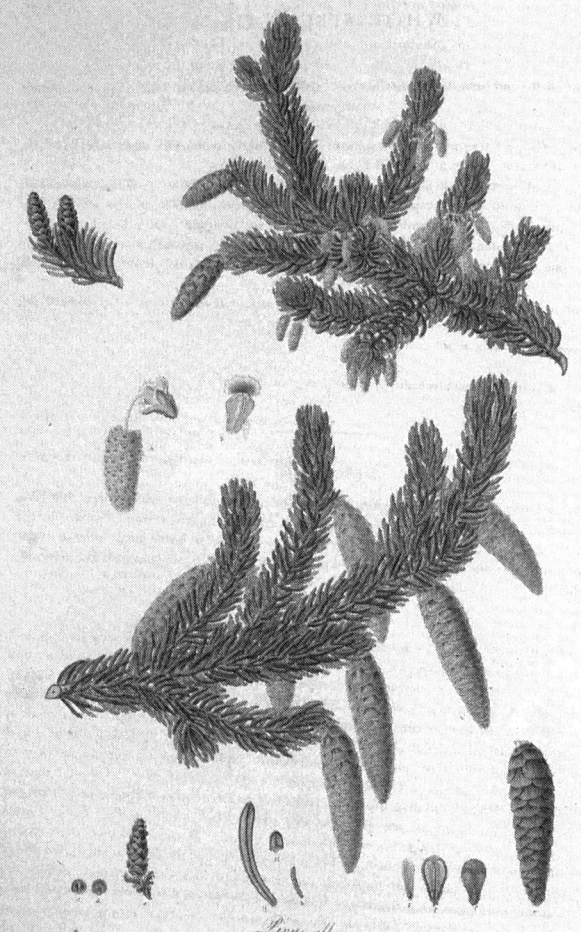

Pinus alba

TAB. 26.

20. PINUS ALBA.

WHITE SPRUCE FIR.

PINUS ALBA, foliis solitariis tetragonis incurvis, strobilis subcylindraceis laxis; squamis obovatis integerrimis.

P. *alba*, foliis solitariis tetragonis: lateralibus incurvis, ramis subtùs nudiusculis, conis subcylindraceis. *Soland. MSS. Ait. Kew. v.* 3. 371. *Willden. Berl. Baumz.* 221.

P. *laxa*, ramulis glaberrimis; phyllophoris elevatis, patentibus; foliis solitariis, sessilibus, subsecundis, tetragonis, obtusiusculis, lineis quatuor longitudinalibus punctatis; strobilis oblongo-ovalibus, pendulis; squamis obovato-subrotundis, integerrimis, tenuibus, lævigatis. *Ehrh. Beitr. v.* 3. 24.

P. *canadensis*, foliis solitariis subulatis bifariam versis, ramulis glabris, cicatricibus sub foliis decurrentibus, conis ovato-oblongis pendulis laxis, squamis subrotundis. *Du Roi. Harbk.* 124. *Wangenh. Beyt.* 5. *t.* 1. *f.* 2.

Abies piceæ foliis brevioribus, conis parvis biuncialibus laxis. *Hort. Angl.* 2. *t.* 1. *Duham. Arb. v.* 1. 3. *n.* 8.

A. canadensis. *Mill. Dict. n.* 4.

Habitat in Americâ Septentrionali.
Fl. Maio.

DESCRIPTIO.

Arbor magnitudine et formâ prioris, at cortice albidiore, foliisque magis incurvatis parùmque minoribus. *Amenta mascula* cernua, longiùs pedunculata, pedunculo gracili; antherarum crista reniformis, dentata, antherâ latior: *fœminea* ovato-cylindracea, erecta; bracteolæ squamis interstinctæ, minimæ, rotundatæ, muticæ. *Strobili* penduli, biunciales, ovato-cylindracei, læves, squamis obovatis, subretusis, integerrimis.

My specimens were procured from the Royal Gardens at Kew. *P. alba* has its name from its *bark* being whiter than that of other species. It is found in America from the forty-third degree of north latitude northward, but farther southward it disappears, requiring a very cold climate. In Canada, Nova Scotia, and the northern parts of New England, it grows in perfection, as Wangenheim informs us, and covers the tops of mountains too bleak and of too bad a soil for *P. Strobus*. The growth of *P. alba* is nearly equal to that of *P. Abies*. It flourishes on poor and rocky land, and also on gravel when dry, and mixed with clay, and a little good mould. A soil apparently but just sufficient to hold the roots enables it to grow. The *flowers* appear towards the end of May. The *cones* ripen at the end of October. These are from two inches and a half to three inches in length, and almost one inch in diameter. The *scales* are smooth, loose, and contain black winged seeds. The *root* commonly sends forth horizontal shoots; but sparingly in light ground. Wangenheim particularly recommends the cultivation of the white Spruce, because it becomes a tree of the first magnitude, the timber of which may be very advantageously employed, and because situations which are unfavourable to the progress of many other pines yield this best. It is much to be wished that advice, founded on arguments so rational, may be generally followed. We are to consider also that *P. alba* is one of the most

x

ornamental of the *Abies* tribe. It grows with its branches feathered down to the ground, and the leaves have a peculiar glaucous hue, making a most beautiful appearance, particularly when mixed with other Pines. The seeds might be procured from Nova Scotia or Canada. When exported from thence, they are usually taken out of their cones, with or without their *alæ* and packed in well-pitched casks. The preparation of the ground intended for the reception of the seed consists merely in digging it up in the preceding autumn. They should be laid on it towards the end of April, without being covered, and pretty thickly, in order that the young plants may not be choaked with weeds when they appear. The attention necessary to the seedling, until it is planted out, is the same as the other species of this genus require. Though the coldest parts of mountains are best suited to the support of *P. alba*, yet experience shews that there is difficulty in raising trees from the seed, and although they come up, and look well the first year, yet they are often lost in the second or third cold winter. Hence it is much better to plant young trees where you wish to fix the plantations. They may be obtained at a very moderate price in almost any of the nursery gardens about London. As the *Abies* tribe seldom or never grow with tap roots, these are fittest for being transplanted, and the age of four or five years seems to be the most proper period. In situations much exposed to cold winds, they should be placed near to one another. When once the plantations begin to thrive, any open spots in them may be sown with success. There are many heaths and waste lands in this and in the sister kingdom, which we may hope to see applied to the cultivation of this species. In England, as proofs that Hounslow and Bagshot Heaths are not unsuitable, I need only mention the flourishing plantations of Whitton and its vicinity. In Ireland, I am certain that the high and mountainous heaths which lie between Westport and the *Killeries*, and the western parts of the County of Mayo [a] might be planted with great advantage, at present they produce little or no profit to their owners.

The bark of *P. alba* is used for tanning; its resin is converted into good turpentine, to which purpose, in those parts of America whence timber cannot be exported, the trees are very generally applied. In Canada, Nova Scotia, and New England, they make another advantage of them, besides using the wood, which is that of preparing *Spruce* or *Essence of Spruce*. This article is exported to the more southern provinces and to England. The mode of preparing it will be described hereafter.

Some of the finest trees of this species, and the greatest number in any one plantation, that I have ever seen, are at Milton-Abbas in Dorsetshire, the seat of the Earl of Dorchester. There are very flourishing ones also at Pain's Hill.

I have remarked that *P. alba*, when young, and in great vigour, will sometimes bear cones of a very large size, with numerous small *squamæ*, and a branch or shoot growing out of their tops.

There is a fine Tree of this species in the beautiful grounds of the Earl of Tankerville at Walton.

EXPLANATION OF TAB. 26.

A.	Male Catkin.
B.	Anthera.
c.	Female Catkin.
d, d.	Its Scales.
e.	Ripe Cone.
f, f.	Scales of the same.
g.	Seed.
h, H, H.	Leaves and their point.

[a] In this county I have observed extensive remains of *P. sylvestris*; the old roots almost covering the bogs, and so little decayed that the poor people dig them up in order to convert them into ropes for tying up the bedding, &c. in their cabins. The situation of these cabins being very damp, ropes made with hemp soon decay. I have bought three fir ropes as thick as cart ropes, and some yards long, in Castlebar market, for sixpence each. The wood is daily cried for sale in the streets of Dublin by the name of Bog Wood.

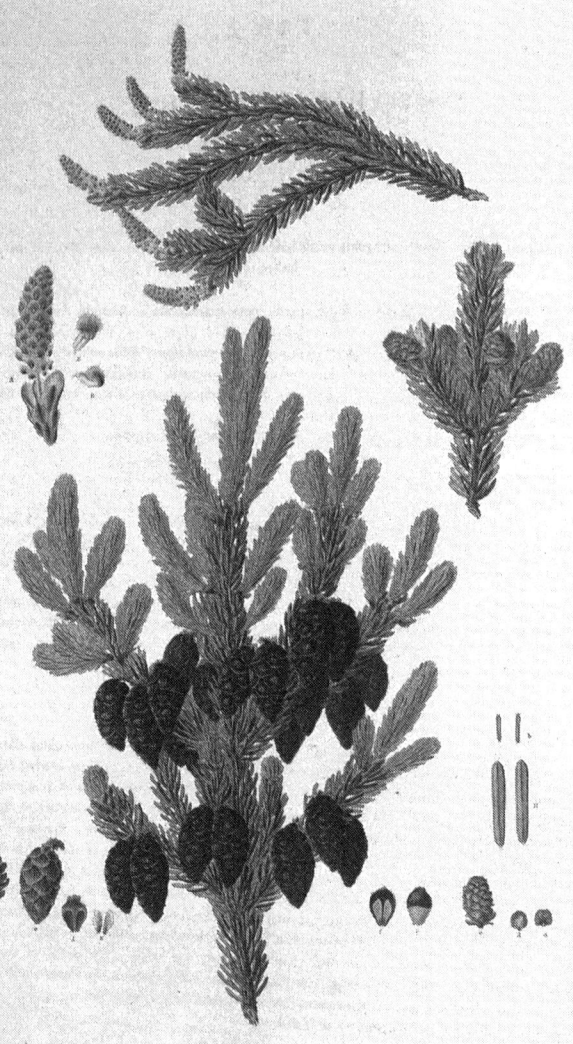

Tab. XXVII

Pinus nigra

TAB. 27.

21. PINUS NIGRA.

BLACK SPRUCE FIR.

PINUS NIGRA, foliis solitariis tetragonis rectis strictis, strobilis ovatis; squamis ellipticis margine undulatis erosis.

P. *nigra*, foliis solitariis tetragonis undique sparsis rectis strictis, conis oblongis. *Ait. Kew. v.* 3. 370.
Willden. Berl. Baum. 220.

P. *mariana*, ramulis pubescentibus, phyllophoris elevatis, patentibus; foliis solitariis, sessilibus, subsecundis, tetragonis, lineis quatuor longitudinalibus punctatis; strobilis ovatis pendulis; squamis
obovatis, crassis, lignosis, rigidis, apice crenulatis, subundulatis. *Ehrh. Beitr. v.* 3. 23.

P. nigra. *Du Roi. Harbk. ed. Pott. v.* 2. 182.

Abies mariana, foliis linearibus acutis, conis minimis. *Wangenh. Beyt.* 75.

Habitat in Americâ Septentrionali.
Floret Maio.

DESCRIPTIO.

Arbor mediocris, erecta, cortice nigricante. *Folia* recta. *Amenta mascula* pedunculata, erecta; antheræ
angustatæ, cristâ rotundatâ, ciliato-dentatâ, antheris duplô latiore: *fœminea* ovata, erecta, bracteolis
exiguis, rotundatis. *Strobili* penduli, ovati, unciales, purpureo-nigricantes, læves, squamis subel-
lipticis, apice dentato-erosis.

P. *nigra* grows wild only in New England, Canada, Nova Scotia, and the colder climates northward. It generally occupies a cold, moist, sandy soil. Its height in such places is from thirty to forty
feet, and its diameter from one to two at about the middle of the trunk, which is uneven, and the
branches reach to the ground. The *bark*, both of the trunk and the branches, is blackish, but the wood
has a reddish white colour; the latter is light, and full of large veins. In cases of necessity, it is used
by the Americans for building fishing-boats, and small vessels. The top of the tree is impregnated
with fine resinous particles. It is at its greatest strength in the spring, when an extract is made from
the leaves and young *shoots*, as well as from those of P. *alba*, with which Spruce beer is brewed.
Some persons are of opinion that the extract made from the former species, is better than that made
from the latter. The *leaves* are little more than half an inch in length, slender, tetragonal, and of a dark
green colour. The *cones* assume an oval shape, but they are small. The *scales* of these are of a
coriaceous texture, and large in proportion to the dimensions of the cone. In Canada and Nova Scotia
the seed ripens about the end of November, but is not shed before the following spring. This tree is
not so much cultivated in this country as it deserves.

Y

The *Pinus Chinensis*, of Professor Pott, (specimens of which he has obligingly presented to me) appears to be no other than the species here described, which indeed seems to be suspected by the Professor himself, from his having inserted the name of "*P. nigra?*"

EXPLANATION OF TAB. 27.

The specimens were procured from Kew gardens.

A. Male Catkin.

B, B. Antheræ.

c. Female Catkin.

d, d. Scales.

e, e. Ripe Cones.

f, f, f. Scales of the same.

g. Seeds.

h, H. Leaves.

Tab XXVII.

TAB. 28.

22. PINUS RUBRA.

NEWFOUNDLAND RED PINE.

PINUS RUBRA, foliis solitariis subulatis acuminatis, strobilis oblongis obtusis; squamis rotundatis sub-
bilobis margine integris.

P. Americana rubra, foliis solitariis subulatis, apice acuminatis, bifariam versis, conis ovalibus pendulis.
Wangenh. Beyt. 75. *t.* 16. *f.* 54.
Newfoundland red Spruce Fir. *Du Roi. Harbk. ed. Pott. v.* 2. 182.

Habitat in Americâ Septentrionali.
Floret Maio.

DESCRIPTIO.

Praecedente humilior, cortice rubro-fusco. *Folia* acuminata. *Amenta mascula* nondum vidi matura:
fœminea ovata sessilia. *Strobili* ovato-cylindracei, unciales vel sesquiunciales, penduli, rubicundi,
læves, squamis cuneatis, apice rotundatis, demùm bilobis, margine integris.

THE specimen represented in the plate was taken from a young tree in the gardens of Messrs.
Whately and Barrett, at Brompton, where it was planted by Mr. Thoburn, who took more pains in
the cultivating of Pines than any gardener of his time. The two full-grown cones at the bottom of the
plate are taken from fine specimens received from America by that able and well known gardener
Mr. Loddige of Hackney. Wangenheim says that *P. rubra* grows only in the more northern parts of
America, and mostly in Nova Scotia and Newfoundland. It is found in a moist cold soil, and never
attains a greater height than thirty feet. The *bark* is of a brownish red colour, smooth on the younger
branches and rough on the older. It is used for building fishing-boats, &c. &c. The *leaves* are not
much more than half an inch in length, awl-shaped and acute, but otherwise resembling those of the
former. These and the young sprouts are used for making Spruce Beer. It is said that the flowers
appear in Nova Scotia towards the end of May. The *seed* is smaller than that of *P. sylvestris*, and ripens
in November. It is contained in oval *cones* which are about one inch or more in length, blunt, and
of a reddish brown colour. They seem to differ from those of *P. alba*, in being rather thicker and the
scales of firmer texture. The *scales* have all a deep notch, and are longer than those of the preceding
species, as well as of a redder colour, which circumstances, added to the more diminutive size of the
tree, distinguish *P. rubra* at once, when it is seen growing with *P. nigra*. Resin runs in abundance out
of the scales when ripe, and overspreads them with a crust, which nature seems to have given by way
of protection against the coldness and humidity of the American winters.

There are a few trees of this species in the nursery at Brompton, but they are too young to produce full-sized cones. I was favoured by Mr. Loddige of Hackney, with *two parcels of cones from America*, one under the name of *P. nigra*, and the other under that of *P. rubra*; the former resembling exactly such as are produced by the trees of that species in England; the latter very different, being larger, longer, more obtuse, and of a shining reddish brown colour; the scales semicircular, each divided by a notch in the middle, and with the margins entire.

Although I have made this tree at present a distinct species, I wish to examine all the different parts of fructification, when an opportunity offers.

That remarkable dwarf Fir, which Lord Clanbrassil introduced some years ago into this country, and the parent tree of which is said to grow on the Earl of Moira's estate in Ireland, I should suppose to be a variety of this species.

There is one of these dwarf trees at Spring Grove, the seat of the Right Hon. Sir Joseph Banks, Bart. and another in Mr. Lee's garden at Hammersmith.

EXPLANATION OF TAB. 26.

a.   A Cone of English growth.
b, b.   Ripe Cones imported from America by Mr. Loddige.
c, c.   Their Scales.
d, d.   Seeds.

Tab. 112.

Pinus Orientalis.

TAB. 29.

23. PINUS ORIENTALIS.

ORIENTAL PINE.

PINUS ORIENTALIS, foliis solitariis tetragonis, strobilis ovato-cylindraceis: squamis rhombeis.

P. orientalis. *Linn. Sp. Pl.* 1421. *Syst. ed. Reich. v.* 4. 178. *Vitm. Sp. Pl. v.* 5. 346.
Abies orientalis, folio brevi et tetragono, fructu minimo deorsum inflexo. *Tournef. Cor.* 41. *Duhamel.*
 Arb. v. 1. 4. *n.* 10.
Ελάτη Græcorum recentiorum. *Tournef.*
Die Morgenländische Tanne. *Linn. Pfl. Syst. v.* 2. 370.

 Habitat in Oriente.

DESCRIPTIO.

Folia brevia, recta, mutica. *Strobili* biunciales, ovato-cylindracei, penduli, squamis cuneato-rhombeis,
 integris.

 I INSERT this species on the authority of Tournefort only, who states, *see Voy. du Levant. tom.* 2. 238,
that he found it growing in the vicinity of Trebisonde, where it is known by the name of Ελάτη. Its
trunk and branches, he says, are about the size of *P. Picea.* The *leaves* are but four or five lines in
length, and not more than half a line in breadth, their colour a shining greenish brown; the *cones* are
described as being nearly cylindrical, about two inches and a half in length, and eight or nine lines in
diameter, pointed, and composed of soft, thin, rounded *scales* which cover very minute and oily seeds.
 I have never seen a specimen of *P. orientalis* either recent or dried, but am inclined to think that
some cones brought from China belong to this species. These cones I have figured; and having been
fortunate enough to obtain a copy of the drawing of *P. orientalis* made by Aubriet under the eye of
Tournefort himself, and which is now in the possession of M. de Jussieu, I am enabled to shew exactly
what that celebrated traveller described. The copy was made by M. Marechal, painter to the museum
at Paris, whose talents are well known; and it was obligingly communicated to me by that eminent
naturalist M. Latreille.

EXPLANATION OF TAB. 29.

 a. Figure of P. orientalis from the original drawing of Aubriet.
 b, b. Cones from China, supposed the same species.
 c, c. Scales.
 d. Seeds.

TAB. 30.

24. PINUS PICEA.

SILVER FIR.

PINUS PICEA, foliis solitariis planis subsecundis, strobilis cylindraceis erectis, bracteolis elongatis, antherarum cristâ bicorni.

P. *picea*, foliis solitariis planis emarginatis pectinatis, squamis coni obtusissimis, adpressis. *Soland. MSS.*
  Ait. Kew. v. 3. 370. *Willden. Berl. Baumz.* 217.
P. *picea*, foliis solitariis emarginatis. *Linn. Sp. Pl.* 1420. *Syst. ed. Reich. v.* 4. 175. *Huds. Angl.* 423.
  Scop. Carn. n. 1193. *Scholl. Barb. n.* 783. *Evel. Sylv. ed. Hunt.* 266. *Pollich. Pall. n.* 914. *Trew.*
  in Nov. Act. A. N. C. vol. 3. *opp.* 445. *tab.* 13. *f.* 29. 44. *Mattusch. Fl. Sib.* 704. *Pall. Fl. Ross.*
  v. 1. 7. *t.* 1. *f. F. Allion. Fl. Ped. v.* 2. 179. *Woodv. Med. Bot.* 575. *t.* 209. *Vitm. Sp. Plant.*
  v. 5. 345. *Villars. Dauph. v.* 3. 809.
P. foliis solitariis, planis, pectinatis, emarginatis. *Hall. Helv. n.* 1657.
P. *Abies*, foliis solitariis emarginatis, conis oblongis erectis, squamis subrotundis planis basi acuminatis.
  Du Roi. Harbk. ed. Pott. v. 2. 133. *Reiter. und. Abel. Abb. t.* 98.
Abies *alba*, foliis subtùs argenteis emarginatis, conis erectis. *Mill. Dict. n.* 1.
A. foliis solitariis apice emarginatis. *Linn. Hort. Cliff.* 449. *Roy. Lugdb.* 89. *Gmel. Sib. v.* 1. 176.
A. Taxi folio, fructu sursùm spectante. *Tournef. Inst.* 585. *Duhamel. Arb. v.* 1. 3. *n.* 1. *t.* 1.
A. conis sursùm spectantibus, sive mas. *Bauh. Pin.* 505.
A. femina 1. Elate telega. *Bauh. Hist. v.* 1. *p.* 2. 231.
Die *Weistanne. Linn. Pfl. Syst. v.* 2. 363.

Habitat in alpibus Helvetiæ, Suæviæ, Bavariæ, &c.
Fl. Maio.

DESCRIPTIO.

Arbor recta, formosa, cortice albicante, lævi, ramis horizontalibus. *Folia* plana, linearia, glaberrima, subtùs glauca. *Amenta* axillaria, cylindracea, obtusa: *mascula* semiuncialia, pedunculata; antheræ basi latiores, cristâ infernè reniformi, supernè bicorni, mucronulis divaricatis: *fœminea* masculis quintuplò majora, bracteolis obcordatis, mucronatis, squamas longè superantibus. *Strobili* erecti, sessiles, cylindracei, ferè spithamæi, bracteolis persistentibus, porrectis, undique muricati, squamis obtusissimis, apice integerrimis, lateribus dentato-ciliatis.

In Siberia, where this species is very abundant, it seems to delight in flat aqueous situations, so much so, that a forest of Silver Firs, as Gmelin states, is considered by the Tartar hordes as a sure indication of good springs being at hand. It grows to a considerable height, and upright, and has a handsome appearance. The *bark* is whitish, and smooth. The *wood* is rather soft, and therefore does not last long, if exposed to the open air. The *branches* shoot horizontally, but are not very numerous,

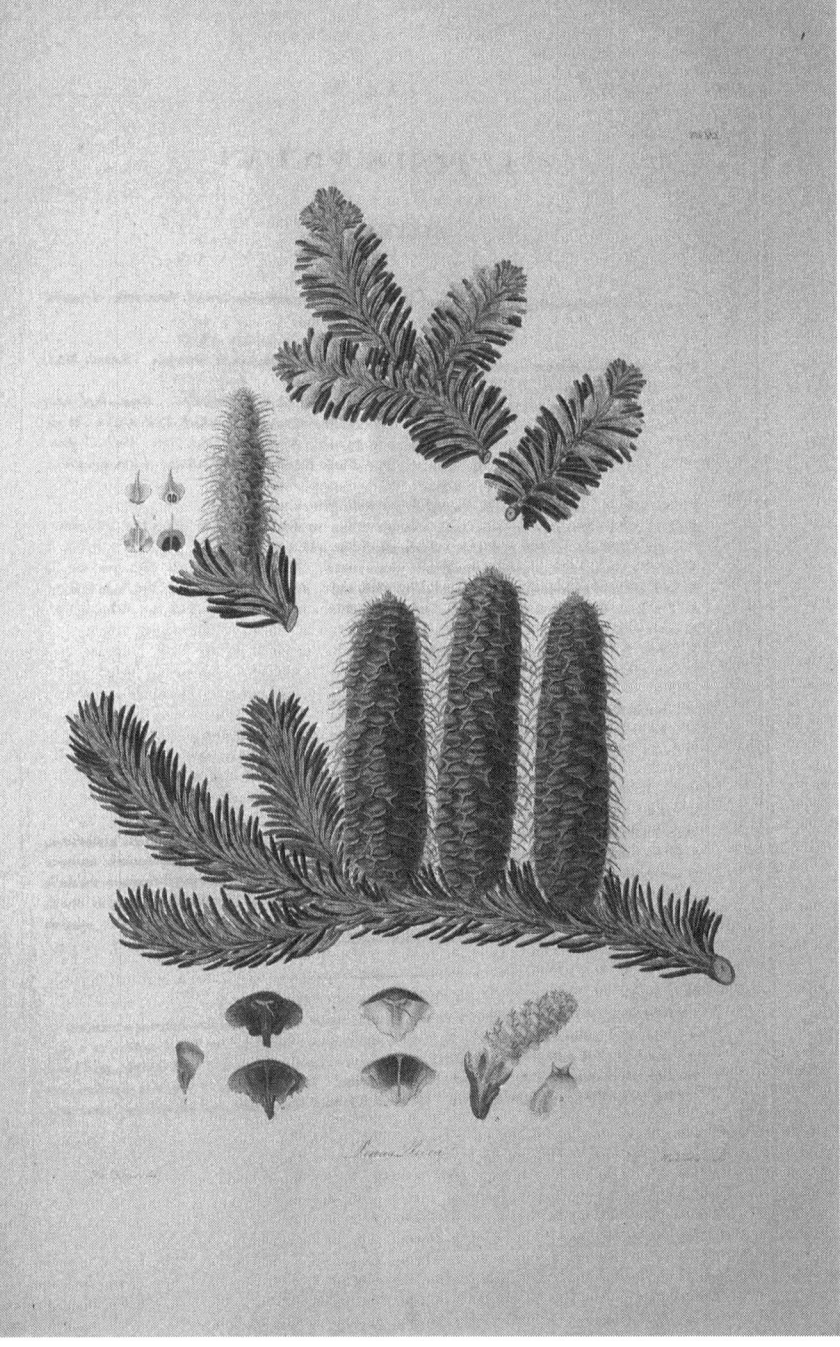

and upon these the *leaves* grow singly. Between the midrib and the edges of the latter, there is a beautiful silvery white appearance, whence the name of *Silver Fir*. They are of a fine strong green colour on the upper surface. The ends are slightly indented. The *cones*, which are of an oblong, or almost of a cylindrical shape, stand erect. The *scales* are roundish, broad and reflexly mucronated. Warm weather soon exposes the *seeds*, which in colour approach the carnation, and in shape are oblong and polygonal. They are wholly covered by the wing on one side, but only half on the other; these wings are obliquely truncated at the top.

The specific name of this tree indicates its yielding Pitch, which is extracted by means of incisions made in the bark, as will be more fully treated of in another place. Haller calls the resin of *P. picea, Terebinthinarum optima*, but the finest turpentine used in our shops seems to be the produce of a very different tree; viz. *Pistachia Terebinthus*.

One of the tallest and finest trees of this species that I have seen, is in the garden of the late John Duke of Argyle, now the property of Mr. Gostling.

EXPLANATION OF TAB. 30.

A.   Male Catkin.
B.   Anthera.
c.   Female Catkin.
d.   Scales of the same, with the prominent bracteolæ.
e, e.   Scales of the ripe Cone.
f.   Seed.

TAB. 31.

25. PINUS BALSAMEA.

BALM OF GILEAD FIR.

Pinus Balsamea, foliis solitariis planis subsecundis, strobilis cylindraceis erectis, bracteolis abbreviatis, antherarum cristâ muticâ.

P. *Balsamea*, foliis solitariis planis emarginatis subpectinatis suprà suberectis, squamis coni florentis acuminatis reflexis. *Soland. MSS. Ait. Kew. v. 3. 370. Willden. Berl. Baumz. 218.*

P. *Balsamea*, foliis solitariis, subemarginatis; subtùs lineâ duplici punctata. *Linn. Sp. Pl.* 1421. *Syst. ed. Reich. v. 4.* 176. *Gron. Virg.* 2. 152. *Wangenh. Beyt.* 40.

P. foliis solitariis subemarginatis, conis ovato-oblongis erectis, squamis subrotundis planis basi acuminatis. *Du Roi. Harbk. ed. Pott. v.* 2. 144.

P. Abies balsamea. *Marsh. Arb. Am.* 102.

Abies *balsamea*, foliis subtùs argenteis apice subemarginatis bifariam versis. *Mill. Dict. n.* 3.

A. taxi folio, fructu rotundiori obtuso. *Hort. Angl.* 2. 2.

A. taxi folio, odore Balsami Giliadensis. *Du Hamel. Arb. v.* 1. 3. *n.* 3.

A. minor pectinatis foliis, virginiana, conis parvis subrotundis. *Pluk. Alm.* 2. *t.* 121. *f.* 1

Die *Balsamtanne. Linn. Pfl. Syst. v.* 2. 365.

Habitat in Virginia, Canada.

Floret Maio.

———

DESCRIPTIO.

Forma ferè præcedentis, at *folia* paululùm angustiora, minùsque glauca. *Amenta* ovata: *mascula* semiuncialia, pedunculata; antherarum crista reniformis, apice mutica, vel brevissimè mùcronulata, nequaquàm bicornis: *fœminea* sesquiuncialia, bracteolis ellipticis, crenulatis, mucronulatis. *Strobili* ovato-cylindracei, violacei, resinosi, fragiles, magnitudine prioris, bracteolis persistentibus, vix squamas excedentibus.

———

P. *Balsamea* has its natural abode in the northern provinces of America, but chiefly in Nova Scotia, Canada, the more northern parts of New York Province, and New England. It stands mostly on the colder side of the mountains, in heavy grounds mixed with clay and sand, yet dry and poor. In these situations we are informed by Wangenheim it grows to a considerable height and strength, like P. *Picea.* Therefore if some trials of this species in England have failed, the principal cause must be either the too great warmth of the spot, or the richness of the soil. I have observed that it bears being transplanted much better than many others of the tribe. The *bark* is of a whitish grey colour, and in texture pretty smooth. Between it and the wood are vesicles which contain a resin, like turpentine,

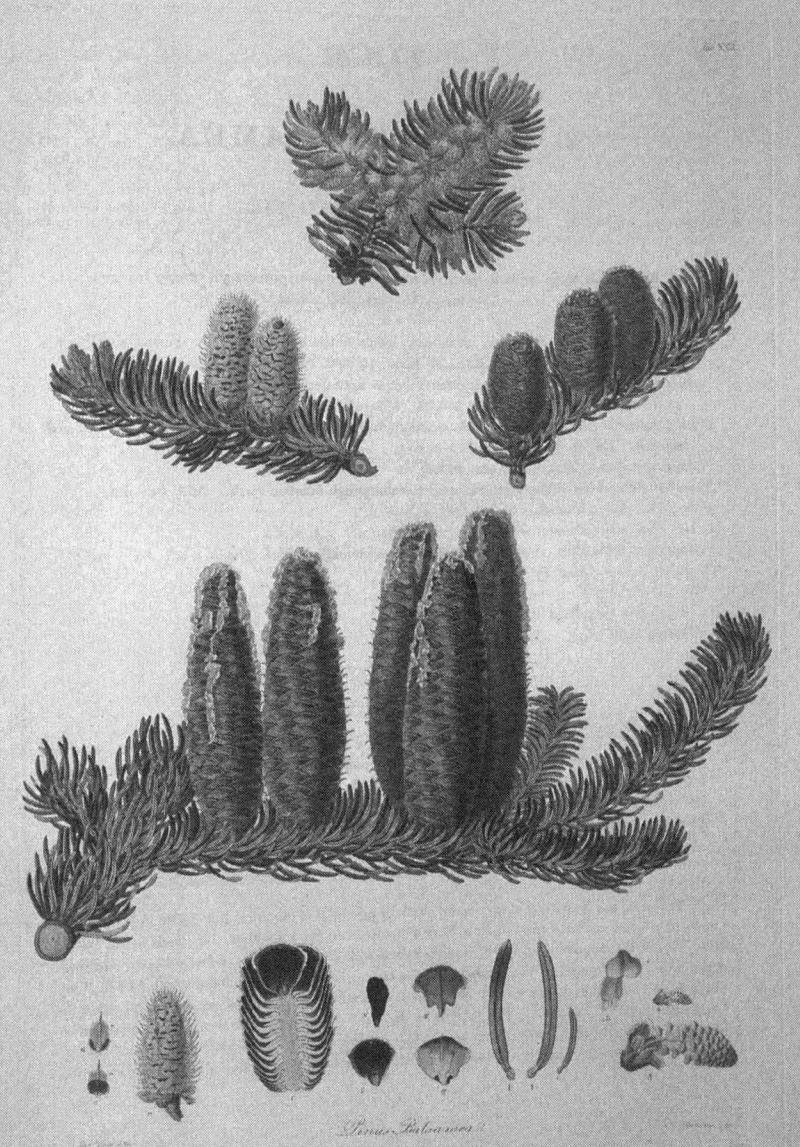

Pl. VII.

Pinus Balsamea.

that is often sold in the shops under the name of Balm of Gilead, though the latter in its genuine state is the produce exclusively of *Amyris Gileadensis*. The resin of *P. Balsamea* is no other than the common *Canada Balsam*. The wood of this species is white, and seems to be better suited to ship-building, and other purposes, than that of *P. Picea*. The *leaves* are somewhat smaller than those of the last-mentioned Pine. The *blossoms* appear at the beginning of May, and the seed ripens about the beginning of October. The *cones* are of a most beautiful glossy deep purple colour, inclining to black, and there exudes from them great quantities of a transparent resin, as is represented in the plate, and which has a very rich appearance. Some of the largest trees of this species, I am informed, are at Wooburn, the seat of the Duke of Bedford, and at Warwick castle, the seat of the earl of Warwick, where they are said to be considerably more than twenty years old, contrary to what was supposed both by Miller and myself, who have observed that they do not last so long in many situations. It does not thrive well in the neighbourhood of London, not growing to any large size, and soon decaying after it is removed out of the nursery. My specimens were procured at Longleat, Wilts, the seat of the Marquis of Bath, the only spot where I have seen this tree in perfection.

EXPLANATION OF TAB. 31.

a, A.   Male Catkin.
B.   Anthera.
c.   Female Catkin.
d, d.   Its Scales.
e, e.   Bracteolæ.
f.   Section of a Cone.
g, g, g. Scales of the same.
h.   Seed.
i, I, I. Leaves.

TAB. 32.

26. PINUS CANADENSIS.

CANADA PINE.

PINUS CANADENSIS, foliis solitariis planis denticulatis subdistichis, strobilis ovatis terminalibus vix folio longioribus.

P. canadensis. *Linn. Sp. Pl.* 1421. *Soland. MSS. Ait. Kew. v.* 3. 370. *Du Roi. Harbk. ed. Pott.* 2. 151.
P. Abies americana. *Marsh. Arb. Amer.* 103.

Habitat in Canada.
Fl. Maio.

DESCRIPTIO.

Arbor vasta, patula, facie Taxi bacciferæ. *Folia* linearia, plana, tenuia, brevia, obsoletè denticulata, subtùs glaucescentia, subdisticha. *Amenta mascula* axillaria, pauciflora, brevissima, et ferè capitata, at longiùs pedunculata; antherarum crista reniformis, apice mucronulata: *fœminea* terminalia, solitaria, ovata, acuta, bracteolis obsoletis. *Strobili* haud unciales, ovati, acuti, læves, squamis paucioribus, rotundatis, integerrimis.

P. canadensis bears a great resemblance to the common Yew, but it is less stiff in its habit, and therefore much handsomer. The *trunk* acquires large dimensions in its native country, but this species has not been cultivated long enough to arrive at any great size in England. The *cones* are oval, pointed, and about the size of a common hazel nut.

This species is not so often met with as might be expected, but I have frequently seen it in the plantations of the curious. It is to be lamented that so elegant a tree is not more frequent in pleasure grounds, especially as it will succeed so well in this country. *P. canadensis* was first cultivated amongst us by the late Mr. Peter Collinson, at Mill Hill, where a large tree is still remaining of his planting.

EXPLANATION OF TAB. 32.

The specimen in the plate was procured from the Royal Gardens at Kew.

a, A. Male Catkin.
B, B. Antheræ.
c. Ripe Cone.
d, d. Scales of the same.
e. Seed.
f, F. Leaves.

Pinus canadensis

Tab. XXIII.

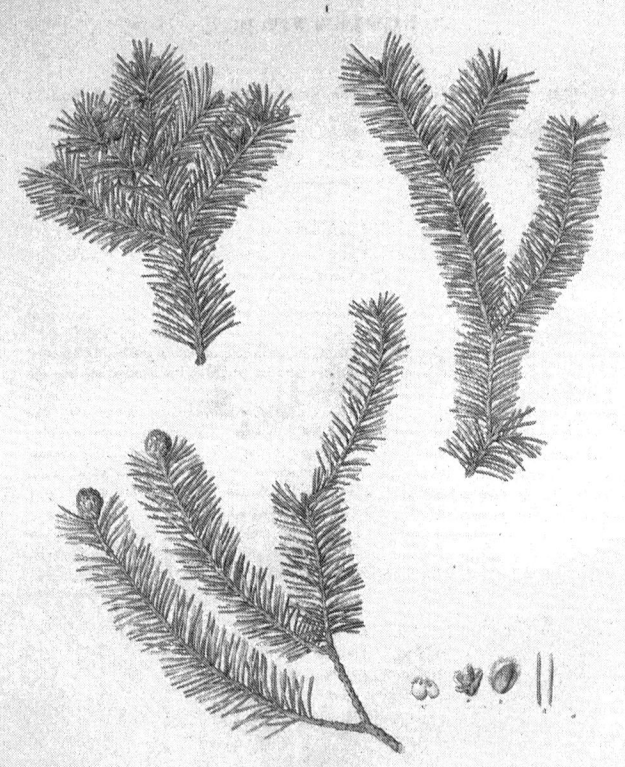

Pinus bifolia

TAB. 33.

27. PINUS TAXIFOLIA.

NOOTKA FIR.

PINUS TAXIFOLIA, foliis solitariis planis integerrimis, strobilis oblongis, antheris inflato-didymis.

Habitat ad Americæ borealis oras occidentales.

———————

DESCRIPTIO.

Habitus P. canadensis, at *folia* angustiora et paululùm longiora, integerrima. *Amenta mascula* ovata, subsessilia, multiflora; antheris inflato-didymis, cristâ reflexâ, minimâ.

———————

THE figure was taken from a specimen in the Banksian herbarium, brought home by Mr. Menzies, by whom it was discovered on the North-west coast of America, and who has favoured me with the following particulars respecting this species.

In general habit this tree resembles *P. canadensis*, and attains considerable height and size. The *leaves* are also very like those of the species just mentioned, but narrower, and their edges are entire, whereas the others are visibly serrated. The *inflorescentia* is much larger than in *P. canadensis* and there are more antheræ. As for the *Cones*, I can give no account of them, those which were brought by Mr. Menzies having been unfortunately mislaid. That gentleman however informs me that they differ in their form from the cones of *P. canadensis*, and that they are longer.

EXPLANATION OF TAB. 33.

a.   Male Catkin.
B.   Bractea.
C.   Anthera.
d.   Leaves.

TAB. 34.

28. PINUS LANCEOLATA.

BROAD-LEAVED FIR.

PINUS LANCEOLATA, foliis solitariis lanceolatis planis patentibus, strobilis globosis; squamis acuminatis.

Abies major sinensis, pectinatis Taxi foliis subtùs cæsiis, conis grandioribus, sursùm rigentibus, foliorum et squamarum apiculis spinosis.   *Pluk. Amalth. Bot.* 1. *t.* 351. *fig.* 1.

    Habitat in China.

DESCRIPTIO.

Folia sessilia, undique patenti-deflexa, sesquiuncialia, lanceolata, acuminato-pungentia, rigida, plana, integerrima, margine scabriuscula.   *Amenta* nondum nobis innotuere.   *Strobili* magnitudine nucis Juglandis, sessiles, nutantes, globosi, læves, squamis ovatis, acutis, mucronulatis.

    THE fine specimen from which the figure was taken, was brought to England by Sir George Leonard Staunton, Bart. and is now in the Banksian herbarium.   The Province in which it was seen growing was Chekiang.

    P. lanceolata is so strikingly different from all the other species of this genus, that it can never be confounded with any one of them.   Its large globular cones, and the singularity of their scales, with the broadness of its leaves, at once distinguish it.   The latter have a glaucous hue on the under surface, and terminate, as well as the scales, in bristly points.

    This tree has not yet been introduced into this country, nor have I heard of any attempt being made to raise it from seed.

EXPLANATION OF TAB. 34.

    a, a. Leaves.
    b.   Scale of a Cone.

Tab. XXXII

Pinus lanceolata.

Tab XXIX

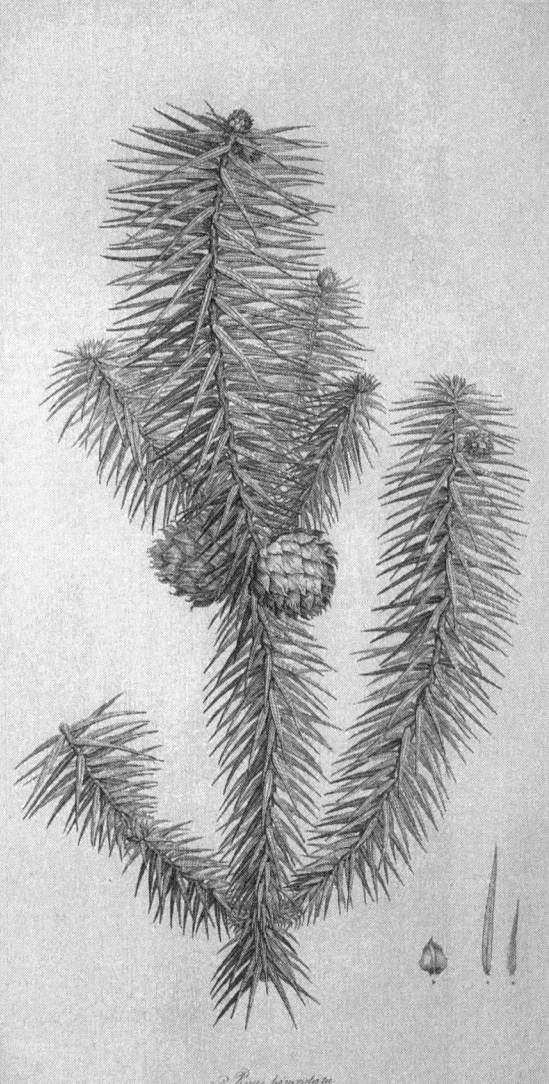

Pinus hamiltonii

Tab. XXIV.

Pinus Larix

TAB. 35.

29. PINUS LARIX.

LARCH.

Pinus Larix, foliis fasciculatis deciduis, strobilis ovato-oblongis, squamarum marginibus reflexis laceris, bracteolis panduriformibus.

P. *Larix*, foliis fasciculatis mollibus obtusiusculis, bracteis extra squamas strobilorum exstantibus. *Soland. MSS. Ait. Kew. v. 3. 369. Willden. Berl. Baumz. 216.*

P. *Larix*, foliis fasciculatis obtusis. *Linn. Sp. Pl.* 1420. *Syst. ed. Reich. v. 4.* 175. *Mat. Med. n.* 472. *Scop. Carn. n.* 1198. *Evel. Sylv. ed. Hunter.* 267. *Trew in Nov. Act. A. N. C. 3. app. t.* 13. *f.* 8. 28. *Pall. Fl. Ross. v.* 1. 1. *t.* 1. *lt. v.* 1. 451, *et* 2. 127. *Mattusch. Fl. Sib.* 703. *Ludw. ect. t.* 86. *Blackw. t.* 477. *Dorr. Nass.* 263. *Allion. Fl. Ped. v.* 2. 178. *Woodv. Med. Bot.* 576. *t.* 210. *Vitm. Sp. Pl. v.* 5. 345. *Villars. Dauph. v.* 3. 807.

P. foliis fasciculatis deciduis. *Hall. Helv. n,* 1658.

P. foliis fasciculatis deciduis, conis ovato-oblongis; squamis ovatis subscabris margine laceris. *Du Roi. Harbk. ed. Pott. v.* 2. 85. *Reitter. und Abel. Abbild. t.* 96.

Abies foliis fasciculatis obtusis. *Linn. Hort. Cliff.* 450. *Gmel. Sib. v.* 1. 176.

Larix *decidua*, foliis deciduis, conis ovato-obtusis. *Mill. Dict. n.* 1.

L. folio deciduo, conifera. *Bauh. Hist. v.* 1. *p.* 2. 265. *Hort. Angl.* 43. *t.* 11. *Duhamel. Arb. v.* 1. 332. *n.* 1. *t.* 1. *Tournef. Inst.* 586.

Larix. *Bauh. Pin.* 493. *Dod. Pempt.* 868. *Cam. Epit.* 45, 46.

Der Lerchenbaum. *Linn. Pfl. Syst. v.* 2. 359.

Habitat in alpibus Helveticis, Vallesiacis, Stiriacis, Carinthiacis, Tridentinis, Sibiricis, &c. Fl. Martio et Aprili.

DESCRIPTIO.

Arbor excelsa, recta, ramis patenti-deflexis, apice pendulis. *Folia* fasciculata, numerosa, undique patula, linearia, obtusiuscula, integerrima, glaberrima, mollia, lætè viridia, decidua. *Vaginæ* breves, crassæ, corrugatæ. *Amenta* lateralia, cylindracea: *mascula* cernua, multiflora; antherarum apice inflato, didymo, cristâ reflexâ, cordatâ, acutâ: *fœminea* erecta, duplò majora, bracteolis exsertis, reflexis, imbricatis, panduriformibus, purpureis, mucronatis, mucrone nervoque viridi. *Strobili* adscendentes, unciâ paulò longiores, ovato-oblongi, fusci, nitidiusculi, squamis rotundatis, striatis, margine aliquantulùm reflexis, emarginatis, et demùm laceris.

2 E

P. Larix is of quick growth, and will rise to the height of fifty feet or more. The *branches* are slender, and their extremities generally hang downwards. They are adorned with long, narrow, soft *leaves*, which spring in tufts from a point, and spread open like the hairs of a painter's brush. Their colour is a light green, and they are deciduous. The *cones* are about one inch in length, obtuse at the *apex*, where they generally assume a purplish colour, and have imbricated *scales*, smooth on their surfaces, but of a lacerated appearance on the edges. Larches are now become extremely common in our nurseries, and it has been remarked that those which have been planted in the worst soils and most exposed situations have thriven best, circumstances which are peculiarly favourable to the increase of their number in this country. They should always be planted in clumps, and at the time of their removal from the nursery, if they be intended for a luxuriancy of growth, should not be more than three or four years old. I am informed that great quantities are now planted in Scotland, where this species is preferred to *P. sylvestris*, on account both of the goodness of the timber, and the quickness of its growth.[a]

The wood of *P. Larix* is used in Switzerland for covering the roofs of houses, being cut into shingles of about one foot square, and half an inch in thickness, which are nailed to the rafters. At first the roofs appear white, but in the course of two or three years become perfectly black, and the joints are stopped by the resin which the sun extracts from the pores of the wood, and which renders the roof impenetrable to rain. The tree is sufficiently frequent in that country to render the covering a cheap one. In Siberia the timber seems to have been used very generally for subterraneous, and even subaquatic purposes, such as the support of vaults, and the repair of canals. Pallas states that some burying places of an unknown nation, and of remote antiquity, still remain with beams and supporters of Larch entire.

By observation made on the strength of timber, it appears, that a beam of Larch, clear and free from knots, and every other imperfection, especially at or near the middle, eleven inches square and six feet and a half long, can bear, if placed horizontally on its two extremities, a weight of two hundred thousand pounds, suspended to the middle of it; and that it can bear a still greater weight in an oblique position.

It is from *P. Larix* that the true Venetian turpentine is extracted. This substance has been procured in the greatest abundance near Lyons in France, and in the valley of St. Martin, near Lucerne in Switzerland. But what is very remarkable, the inner part of the wood of this tree yields a pure gum, scarcely inferior in its qualities to the Arabian gum. In the Russian empire this has been received into the shops, and sold under the name of Orenburgh gum, an appellation extremely improper, as Pallas justly observes, Orenburgh being very distant from the Uralensian forests, where the gum is collected. The largest and handsomest Larch I have ever seen is at Stratfieldsay, the seat of Lord Rivers. The trunk of this tree is six feet in circumference, at the height of four feet and a half from the ground, and in proportion quite to the top. Its branches rest on the ground, extending over a space forty feet in diameter. It was planted between forty and fifty years ago by Mr. Malcolm, nurseryman, of Kennington Common. A peculiar *Boletus* draws its nourishment from *P. Larix*, and hence has acquired the trivial name of *laricinus*.

Miller mentions three varieties of *P. Larix*. The first of which he says is a native of America, and must be our *P. pendula*, and the second, which is said to be brought from Siberia, is probably the variety growing at Sion House alluded to in my description of *P. pendula*. It is difficult to determine what is meant by his third variety, called *Larix Chinensis*, all the trees being now dead.

Since the above was written, that part of Professor Martyn's new edition of Miller's Dictionary, in which the genus *Pinus* is described, has been published, and I find the learned Professor's collections on the subject of the Larch so full and valuable that I beg leave to refer the reader to that work.

---

[a] A pamphlet has lately been published, entitled, " A Treatise on the Cultivation of the Larch, and Scotch Fir Timber, &c." by Mr. Pontey, octavo, which places the advantages attending the culture of this tree, in a strong point of view, and which I recommend to the perusal of those who wish for further information on this subject.

EXPLANATION OF TAB. 35.

a, A.   Male Catkin.
b, B, B. Antheræ.
c, C.   Female Catkin.
d, D.   Scale of the same.
e, E.   Inner Scale or Bracteola.
f.   Unripe Cone.
g, g.   Scales of a ripe Cone, with the Seeds.
h, H.   Leaf.

TAB. 36.

30. PINUS PENDULA.

BLACK LARCH.

PINUS PENDULA, foliis fasciculatis deciduis, strobilis oblongis; squamarum marginibus inflexis, bracteolis panduriformibus acumine attenuato.

P. *pendula*, foliis fasciculatis mollibus obtusiusculis, squamis strobilorum bracteas tegentibus. *Soland.*
MSS. Ait. Kew. v. 3. 369. *Willden. Berl. Baumz.* 215.

P. *intermedia*, foliis fasciculatis deciduis, conis ovato-cylindricis laxis; squamis subrotundis retusis. *Du*
Roi. Harbk. ed. Pott. v. 2. 115. *Wangenh. Beit.* 42. *t.* 16. *f.* 37.

P. Larix nigra. *Marsh. Arb. Am.* 103.

Habitat in Americâ Septentrionali.
Floret Maio.

DESCRIPTIO.

Habitus præcedentis, ramulis magis elongatis et pendulis. *Folia* ferè prioris, sed breviùs vaginata.
Amenta mascula multiflora, antheris minùs inflatis, et breviùs cristatis: *fœminea* cylindracea, obtusa,
bracteolis ut in præcedente. *Strobili* vix unciales, fusci, nitidi, ovato-cylindracei, squamis paucioribus,
margine inflexis, integris.

P. *pendula*, as we are informed by Wangenheim, shews itself only in the cold mountainous parts
of N. America, from the forty-fifth degree of north latitude; in such tracts it grows to a tall strong
tree. Its native soil is a rich clay mixed with sand. The *bark* is of an ash-grey colour, and the
wood reddish. The *branches* are weak and drooping. The *leaf-buds* are almost black, and yield a
fine turpentine. The *leaves*, when full grown, assume a dark green colour, and fall in the autumn.
In the province of New York, under the forty-second degree of north latitude, the *flowers* appear
towards the end of April. The *seed* ripens in September, and is readily shed by the cones, the smooth
scales of which sit very loosely. The *cones* are about half an inch commonly in length, they are
rounded at the base, and of a yellowish brown colour. The *wood* of this tree is said to be good and
durable. The first tree planted in this country was that which grew at Mill Hill in the gardens of
the late Mr. Collinson, and which was the finest and largest tree I have ever seen, bearing great quantities of cones, with ripe seed, annually. It was from that tree I procured the specimen from which the
plate was taken. Professor Pott observes that the trees he has seen growing in Germany are two feet
ten inches in the circumference of the stem, and are fifty feet in height; they do not grow so fast as

Pinus pendula

the White Larch, but they surpass the *P. microcarpa*, and they have besides the advantage of the latter in the very straight growth of their stem, in which they exceed even the White Larch. The wood is firm, and seems likely to be good and useful.

EXPLANATION OF TAB. 36.

a, A.	Male Catkin.
B, B, B.	Antheræ.
c, C.	Female Catkin.
D.	Scale of the same.
E.	Bracteola.
f, f.	Ripe Cone.
g, g.	Scales of the same with the permanent bracteola.
h, h.	Scales with the Seeds.
i.	Seed.
k, K.	Leaves.

TAB. 37.

31. PINUS MICROCARPA.

RED LARCH.

Pinus microcarpa, foliis fasciculatis deciduis, strobilis subrotundis paucifloris; squamis inflexis, bracteolis ellipticis obtusè acuminatis.

P. *laricina*, foliis fasciculatis deciduis, conis subglobosis; squamis laxis orbiculatis glabris. *Du Roi. Harbk. ed. Pott. v. 2. 117.*

DESCRIPTIO.

Præcedentibus similis, *ramulis* laxè pendentibus, *foliis* paululùm minoribus. *Amenta mascula* brevissima et ferè capitata, antheris apice lateribusque ventricosis, cristâ muticâ, deflexâ: *fæminea* retusa, pauciflora, bracteolis obovato-ellipticis, acumine obtusiori. *Strobili* parvi, semiunciales, rubicundi, squamis paucioribus, margine inflexis, integris.

THIS species is very scarce in England, but would be a great ornament to the finest plantations. The only tree of any size I have seen is at Whitton, where it was planted by John Duke of Argyll, and which has a remarkably beautiful appearance in the summer, being covered with a great number of bright purple *cones*. The specimen from which the figure was taken came from that tree. It is a very remarkable species, the cones being much smaller than those of *P. pendula*. Upon examining the two trees very accurately, I am inclined to suppose them really distinct: besides the smallness of the cones they differ essentially in the figure of the *bracteolæ*. The cones of both are sent from America annually to Mr. Loddige, one under the name of the black, and the other of the red Larch. He has a large plantation of fine healthy trees of each sort about eight feet high, which produce many cones every year; and although they grow close to each other, the cones always remain distinct. There are two trees growing at Sion House under the name of the Siberian Larch, which I make no doubt were brought from America, and appear to be *P. microcarpa*.

EXPLANATION OF TAB. 37.

A.   Male Catkin.
B.   Anthera.
C, C.   Female Catkins.
D.   Bracteola.
E.   Inner view of the same.
f.   Scale of the ripe Cone, with the permanent Bracteola.
g.   Scale with the Seeds.
h.   Seeds.
i, I.   Leaf.

Pinus microcarpa

Tab. XXXV N.1.

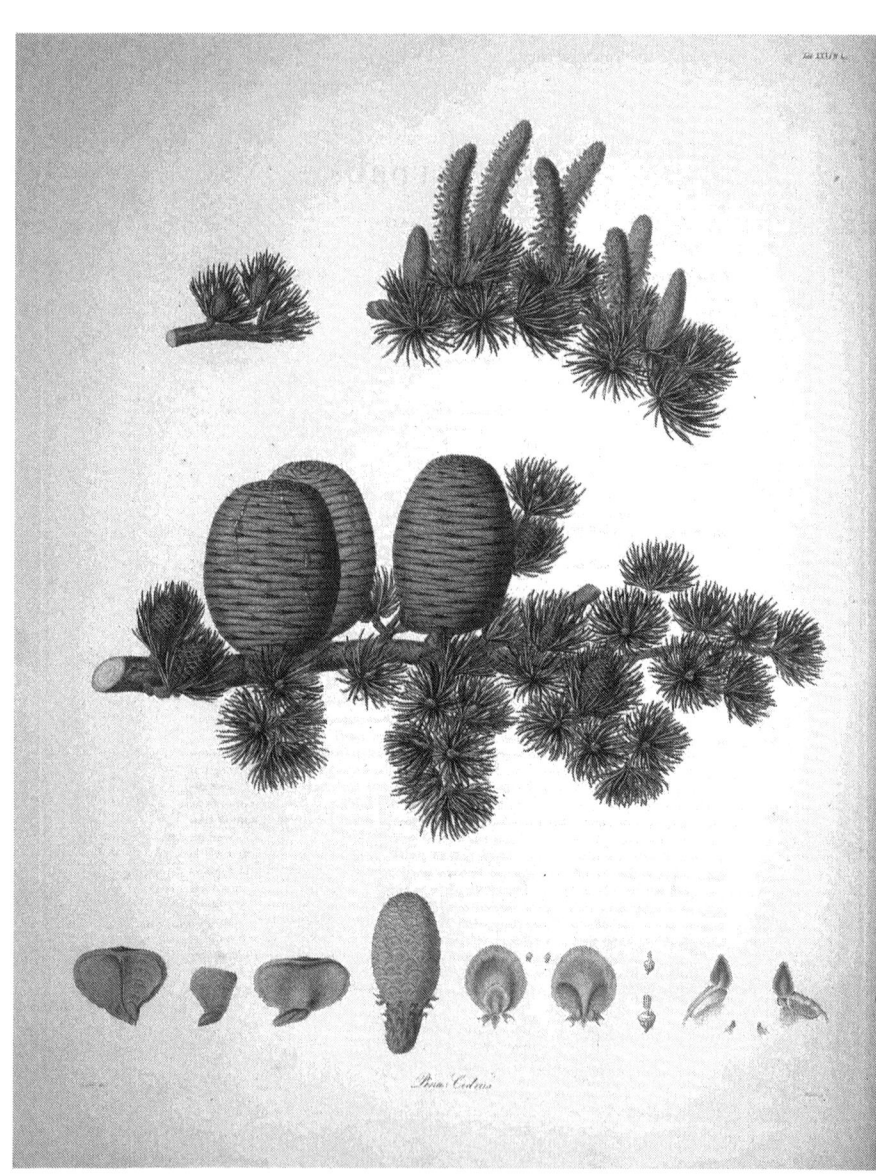

Pinus Cedrus

32. PINUS CEDRUS.

CEDAR OF LEBANON.

PINUS CEDRUS, foliis fasciculatis perennantibus, strobilis ovatis obtusis erectis; squamis adpressis rotundatis.

P. *Cedrus*, foliis fasciculatis acutis.   *Linn. Sp. Pl.* 1420. *Syst. ed. Reich. v.* 4. 174.   *Evel. Sylv. ed.*
   Hunter. 311.   *Ait. Kew. v.* 3. 369.   *Vitm. S. Plant. v.* 5. 345.   *Willden. Berl. Baume.* 214.
P. foliis fasciculatis perennantibus, conis ovatis obtusis erectis; squamis adpressis rotundis, cortice lævi.
   Du Roi. Harbk. ed. Pott. v. 2. 120.
Larix Cedrus.   *Mill. Dict. n.* 3.
L. orientalis, fructu rotundiore obtuso.   *Tourn. Inst.* 586.   *Duhamel. Arb. v.* 1. 332. *n.* 2. *t.* 132.
Cedrus, foliis rigidis acuminatis non deciduis, conis subrotundis selectis.   *Trew. Ehr. t.* 1, 4. 28, 60, 61.
   Nov. Act. A. N. C. v. 3. *App.* 445. *t.* 13. *f.* 1. 7. 11. 12. 14.
C. conifera, foliis laricis.   *Bauh. Pin.* 490.   *Raii. Hist.* 1404.
C. Libani.   *Barr. Ic.* 499.   *Edw. Ornith. t.* 188.
C. Phœnicia.   *Renealm. Sp.* 27.
Cedrus.   *Bell. It.* 162.   *Cam. Epit.* 57.
Die Wahre Ceder.   *Linn. Pfl. Syst. v.* 2. 386.

   Habitat in Syriæ, Libani, Amanæ, Tauri montibus.
   Floret Octobri.

───────

THIS is too remarkable a tree to be mistaken or confounded with other species.   Its *branches* are very long, and disposed like those of *P. Larix*.   After the excision of a branch, the part remaining in the trunk gradually loosens itself, and assumes a round form, resembling a potatoe; if the bark covering it be struck smartly with a hammer, the knot leaps out.   This fact was communicated to me by Sir Joseph Banks, and I have since repeated the experiment myself.   The *leaves* are small, stiff, and of a fragrant smell.   The *cones* are very large, and have rounded, membranous *squamæ*, which in their natural state are pressed close, but unfold when exposed to warm water.   There are six cotyledons. From the observations made by Pallas in Siberia, dry ground is not so favourable as moist to the growth of *P. Cedrus*.   This traveller asserts that a wet season, in the districts which he visited, was considered as absolutely necessary to bring the fructification to maturity.   The seeds should be sown about the middle of March in pots, or boxes, nearly half an inch deep.   The following is the best mode of extricating them from the cones.   Let a hole be bored with a gimblet exactly through the middle of each cone, from the base to the apex.   Put them into a tub of water, in which they may remain until the next day; then let a wooden peg, rather bigger than the gimblet, be thrust into the hole, and it will so divide the cones that the different scales may be taken away, and the seeds picked out.   In this process, great care must be taken *not to bruise the seeds*, as they will be very tender.   The plants will come up in about seven or eight weeks after the seeds have been sown; they should then be removed from the heat of the sun into a shady place, where they may stand, but not under shelter, the whole summer; during which time it is necessary to keep them free from weeds, and to water them occasionally.   In the winter season a warmer situation is to be sought, and if the weather should prove very severe, perhaps it may be proper to shelter them with mats, or to cover them with a hot-bed frame. At the beginning of the following April, these plants may be pricked out into beds, and placed about four inches apart. Should the weather be dry, it is advisable to give them shade and moisture until they have taken root. After having been two years in the beds, they must be transferred to the nursery, where they may remain until the place of their final destination be ready. Whilst the young cedars are in the nursery, and indeed after having been planted out, many will have a tendency to droop, probably in their leading shoot. As soon as this is perceived, an upright stake should be driven into the ground, and the shoots tied to it with matting, to keep them upright. It may not be amiss also, in some instances, to lighten the head by cutting off the extremities of some few of the large branches. When the trees have been finally transplanted however, they should be in general left to nature. Not a knife nor a hatchet should be brought near the old part of the branches, for the lopping the thick wood will not only retard their growth, but injure their beauty. The Cedar is extremely tardy in its increase of size, even under the most favourable circumstances, so that the greatest caution ought to be observed in the rearing it. The epithet of lofty, commonly given to the Cedar, is by no means applicable, since from the accounts given of those which still remain on Mount Lebanon, they are not very high, though their branches spread widely. The last-mentioned circumstance warrants the fine allusion of the Psalmist, in describing a prosperous people. " They shall spread their branches (says he) like the Cedar tree." Of the few trees of this species still standing on Lebanon, seven, we are told by Billardiere, are of amazing size. The trunk of the largest is nine feet in diameter. They are all preserved with religious care, and on the day of the transfiguration, a solemn festival is celebrated on the mount, called the *Feast of Cedars*. The memoirs of the Levant missionaries state that the Patriarch officiates pontifically on this devout occasion, and threatens with ecclesiastical punishment those who may presume to diminish or hurt the Cedars that grow on the consecrated spot.

The diuturnity of the Cedar we frequently find alluded to. The wood of this famous tree has been supposed to preserve books much better than any other material; hence the expression " *Cedro dignus*," was considered as one of the highest compliments that could be bestowed on a literary performance.

The wood was not liable to be corroded by insects, on which account it was much used in ancient times for coffins, and chips of it were considered as destructive to moths and worms. It is recorded that in the temple of Apollo, at Utica, was found Cedar wood nearly two thousand years old, and at Saguntum in Spain, in an oratory consecrated to Diana, two hundred years before the destruction of Troy, a beam was discovered which has since been removed to Zante.[1]

But in the relation of the properties assigned to this tree, I think with Professor Martyn there is much vulgar error and confusion, the Cedar of Lebanon being often confounded with trees which belong to different *genera*. At least the accounts given by the ancients of the long duration of their Cedar, very ill accord with the species now under consideration, whose wood is no more than a very inferior kind of deal, with little or no smell, and of a soft texture, evidently of short duration. This appears by a table in the possession of Sir Joseph Banks, made of the Hillingdon Cedar, one of the largest that ever grew in this country. The word *Cedrus* seems in many cases to be ambiguously used by Greek and Latin authors, but appears in general much better to apply to the *Cupressus horizontalis* of Miller, which I have no doubt is a distinct species from *C. sempervirens* of Linnæus. Perhaps it may not be superfluous here to mention that the wood used for black-lead pencils is not *Pinus Cedrus* but *Juniperus Bermudiana*.

By whom the Cedar was first introduced into England, I have not been able to ascertain. In the Gentleman's Magazine for March 1779, Sir John Cullum has taken great pains to settle this point, and concludes that we are very probably indebted to Mr. Evelyn for its introduction.

Some of the most vigorous and beautiful in this country at present are growing at Pain's Hill, Whitton, and Chiswick. They produce a prodigious number of cones annually.

This species being so fully and accurately illustrated in Ehret's figures, published by Trew, I thought it unnecessary to give a plate or botanical description of it at present.

[1] *Evelyn's Sylva*, 315.

Pinus ....

TAB. 38.

32. PINUS DAMMARA.

AMBOINA PITCH PINE.

Pinus Dammara, foliis oppositis elliptico-lanceolatis striatis.

Dammara alba. *Rumph. Amboin. v.* 2. 174. *t.* 57.
Arbor Javanensis Visci foliis latioribus conjugatis, Dammara alba dicta.   D. Sherard.   *Raii. Hist. v.* 3.
  dendr. 130.   *Herb. Sherard.*

Habitat in Amboinæ excelsis montibus solo argillaceo.

DESCRIPTIO.

Arbor, ex auctoritate Rumphii, abietiformis, excelsa, caudice simplici, tereti, glabro, comâ parvâ.
Ramuli foliosi, tetragoni, glabri. *Folia* opposita, decussata, breviùs petiolata, elliptico-lanceolata,
obtusa, integerrima, coriacea, glaberrima, nitida, nervis plurimis parallelis obsoletè striata. *Strobilus*
ovatus, squamis obtusis, muticis, suprà marginatis. *Semina* elliptica, compressa, sulcata, apice emar-
ginata, hinc alata, alâ rotundato-cuneiformi.

Having but lately become acquainted with this curious species, I am obliged to introduce it as an
appendix to the rest. Its most natural place in the genus is near *P. lanceolata.* For the specimens
represented in the plate I am obliged to Sir Joseph Banks, in whose herbarium the leaves are preserved,
and who has lately received fragments of the cone from Mr. Christopher Smith, Botanist to the East
India Company, at Amboina. Dr. Smith has also discovered a specimen of the leaves in the Sherardian
herbarium at Oxford, among the plants collected by Dampier. From these materials, and the account
given by Rumphius, all our knowledge of this tree is derived. What he has accidentally (as he was
ignorant of the sexes of plants) called the male and female trees, appear to us to be really so, and that
this species is dioecious. His account of the valuable resin for which it is most remarkable is here
subjoined.

The following is an account of the resinous substance produced by this tree, which is well known in
India under the name of *Dammar-Puti, Dammar-Batu,* or *White Dammar,* and has been thus
described by Rumphius, in his *Herbarium Amboinense.* (Lib. 3. cap. 10.)

The pellucid resin which flows from this tree is at first soft and viscous, but within a few days it
becomes as hard as stone, and has all the transparency and whiteness of crystal, especially that which
adheres to the trees, and it will sometimes hang from them in the shape of icicles; that which flows to

the ground, however, becomes black, and mixed with extraneous matter. These *icicles* (as they may be called) are sometimes as much as a palm in breadth and a foot in length, and exhibit an elegant, striated appearance. For the first half year, the resin retains its white colour and transparency, but afterwards assumes a beautiful amber colour. It is brittle, and when broken shines like glass. It is much harder than the sort known by the name of *Dammara Selanica*, and in some degree admits of being bent, but when pounded it is friable. The product of the male trees is more white and pellucid, but dries more slowly and exudes in a smaller quantity, whence little or none is collected from them. To force a supply of this substance, it is usual to make incisions in the lower part of the trunk with sharp knives. This occasions the formation of large knots in the wounded places, which protrude like heads, as in the maple, are covered with dammar, and put forth a number of branches. The peasants clear away the rubbish around the feet of the trees, and dig holes, in order to collect the dammar free from impurities; where the roots rise bare out of the ground, however, they are made knotty by incisions, and yield resin like other wounded parts of the tree, becoming covered with a sort of white bark. The smell of fresh and soft dammar is perfectly resinous, but when dry this substance does not emit any particular odour; thrown on burning coals, it gives out a smell partaking of turpentine and mastich, as does also the taste, though the latter is somewhat like the *Canarium*. It is very inflammable, and burns longer than the *Dammar Selan*, but without any crackling, though it emits a great quantity of acidulous smoke, which produces a very unpleasant effect on those who are unaccustomed to it. As the common dammar, sold in the market, is white and semitransparent, as well as that collected from the knots of the lower part of the trunk, so is the latter changeable in its colour, varying from a reddish to a horn, and even to a black appearance; it is, nevertheless, hard and pellucid, like the great masses and heads which hang from the thicker branches and oldest trees, for, as these cannot on account of their height be ascended, the masses hang on them the longer, and lose their original whiteness and become of a horny colour. This circumstance is particularly remarkable on the dammar trees about *Way*, whence I am led to believe that the variations of colour proceed from the difference of the time of year, or from the interval that the masses remain in their native situations. In the year 1688, I sent a piece of dammar to the University of Leyden, which in its form resembled the head of an infant, and by artificial means had been made to assume something like features, but the nose was very ugly, and there were red marks near it, resembling streaks of blood. I have also in my own possession a large white semitransparent mass, which resembles an immense ox's gall bladder. Some of the crystalline branches sent into Holland did not retain their colour, but became, there, of a sort of amber hue.

The Malay name of *Dammar Puti*, and *Dammar Batu*, signifies *stone resin*, for it is the hardest of all the Dammars, and approaches very near to the *Gum Animæ*. Among the Ternats it is called only *Salo*, or *Salo Bobuda*; in Amboina *Camal Camar*, and *Cama*; about Lariqua *Isse*; and about Grisecca, in Java, *Dama*.

The medicinal uses of this resinous substance have not yet been discovered. Some of the people of Amboina, however, (but I own I would not, myself, be of the number) whose feet have been wounded with thorns, or splinters of wood, have no sooner extracted the latter than they have dropped into the punctured parts a drop or two of burning dammar, which has prevented the formation of an ulcer, and scarcely created any pain in the callous, hard soles of the Indians. This species of Dammar is not easily liquefied, except by means of an admixture of Calapp oil, or common pitch. It is not found, however, to adhere well to the keels of ships, but is apt to fall off, and therefore does not answer the purpose of pitching; yet it deserves further trial, especially in a country where the want of pitch is attended with so much danger to navigation: other resins, indeed, will in some measure supply its place. To persons who write much (as clerks and secretaries) the dammar is of some use, for if they have occasion to scrape out a letter or a word, and will afterwards sprinkle a little of the powder on the place from which the erasure has been made, the paper is rendered smooth and susceptible of being again written on; but the letters soon become faint.

The *Dammara Radja* (called by the Ternaats *Salo Colano*) is the same resin as that above described, but only the largest and most transparent pieces, which hang from the upper part of the trunk, free from impurities. That which is found in Batsjana and Hallemapera never turns yellow like the Amboina dammar, but is almost always white, and very gradually acquires any tinge of yellow; in taste and smell, however, it is the same, and collected from the same kind of tree. This sort alone is employed by the kings of the Moluccas as a suffumitory, whence it obtains the name of *royal dammar*, and the common people are prohibited from using it. There is some amusement in observing people unacquainted with this substance, who, meeting with it in the shops, take it to be lump sugar: as it is sold at a very cheap rate, the purchasers seek out the first corner to enjoy a taste of it, but immediately discover their mistake.

In books of voyages, we sometimes read of large quantities of mastich being found in these islands, but the substance alluded to is no other than the *Dammar Batu*, which, when burnt, emits the same smell."

<hr>

EXPLANATION OF TAB. 38.

a.   Branch of the *Pinus Dammara* from Sir Joseph Banks's herbarium.
b, b.   Scales of a ripe Cone.
c.   Seed with its Wing.
d.   Is supposed by Mr. Francis Bauer, who made the drawing, as well as by myself, to be a female Catkin, and that the oblong bodies at the base of the enlarged scales E, E, may be the real stigmas, which, if true, will throw a new light on the fructification of this genus. Dr. Smith is rather persuaded, from an attentive consideration of the description in Rumphius, that this is the Male Catkin, and that the bodies in question are *Antheræ*.

OF THE

MEDICINAL AND OTHER USES OF VARIOUS SUBSTANCES

PREPARED FROM

TREES OF THE GENUS PINUS,

BY

WILLIAM GEORGE MATON, M.D.

FELLOW OF THE ROYAL COLLEGE OF PHYSICIANS, LONDON, F. R. A. & L. SS. &c.

Most species of Pinus may be made to yield (and many of them produce spontaneously) a remarkable resinous juice, usually called *Turpentine*. This appellation more properly belongs to the product of a different genus called by Linnæus *Pistachia*, which contains the true *Terebinthus* of the ancients. The juice of Pines, however, like that of the Turpentine trees, has an austere, astringent taste, singular viscosity and transparency, ready inflammability, and a disposition to become more or less concrete. In distillation with water, it yields a highly penetrating essential oil, and the liquor is found to be impregnated with an acid, a brittle resinous matter remaining behind. Digestion with rectified spirit of wine completely dissolves all the resinous part, along with which some portion of the insipid gum, or mucilage, is also taken up. If this solution be filtered, and diluted largely with water, it becomes turbid, and throws off the greatest part of the oil, the gummy substance being retained. If the solution be subjected to distillation, the spirit brings over with it some of the lighter oil, so as to be sensibly impregnated with its terebinthinate odour, and it leaves behind an extract differing from the resin separated by water, in having an admixture of mucilage. The native juice becomes miscible with water, by the mediation of the yolk or the white of an egg, but more elegantly by that of vegetable mucilage, and forms a milky liquor. Exposed to the immediate action of fire, the roots and other hard parts of the trees produce a thick, black, empyreumatic fluid, which, containing a proportion of saline and other matter mixed with the resinous and the oily, proves soluble in aqueous liquors, and, according to its several modifications, constitutes the varieties of *Tar* and *Pitch*. The resinous *residua* of the several processes to which the matter extracted from Pines may be subjected constitute the varieties of *Rosin, Colophony*, &c. There are also other products, both native and artificial, much employed in medicine and the arts, and which have correspondent denominations, to be specified in their proper places.

The terms commonly attached to these substances are, in general, extremely vague, ambiguous, and inexpressive. Those employed in ancient authors are not to be excepted from the application of this remark; they have occasioned great difference of opinion among commentators, and, in some instances, they remain to this day undefined; but, on the whole, they were used with more precision perhaps than is observable either in the popular discourse, or in the regular *pharmacopœia*, of modern times. In the following pages, which are intended to describe the several substances and processes in detail, we shall endeavour to dissipate the confusion so far as we are able, by substituting appropriate appellations for those which are either ambiguous or likely to lead to error, and by arranging immediately under every head such synonyms as may be adduced without undue latitude of conjecture.

As so many trees of this genus yield the same substances, and as in different countries different trees have been had recourse to, authors will be found to vary very much in their references to the species of *Pinus*,

---

* The *Terneth* of Theophrastus, (lib. 3. c. 2.) and Dioscorides, (lib. 1. cap. 76.) from which the word *Terebinthus* seems to have been derived. *Pistachia Terebinthus* yields the resinous juice called in the shops *Cyprus* and *Chio* Turpentine, the superiority of which to all the products of the Pine tribe was well known to, and described by, most of the ancient writers on the *Materia Medica*, (See Disc. loco supra citato.) Genuine turpentine is almost colourless, and emits a peculiar odour, much more agreeable than that of the common turpentines of the shops.

from which these substances are respectively procured. The processes themselves are also somewhat different among different nations. We shall content ourselves in this treatise with pointing out those species which have been generally considered as most proper for the several purposes, and, in regard to the processes, confine ourselves to those which appear best suited to their respective objects.

Medicinal Properties of Terebinthinate Substances in general.

Terebinthinate substances, when taken internally, seem to warm the *viscera*, raise the pulse, and impart additional excitement to the whole vascular system; applied externally, they increase the tone of the part, counteract indolence of action, and deterge, as it were, ill-conditioned ulcers. *Internal* ulcerations indeed, especially in the urinary passages, as well as laxities of the seminal and uterine vessels, are supposed to be diminished by the exhibition of preparations of this nature. They certainly seem to act in a peculiar manner on the urinary organs, impregnating the water with a violet smell, even when applied externally.[*] Most of them produce a laxative effect on the bowels, when given in a certain dose. There also appears to be a sort of styptic property in some preparations of turpentine, on which account recourse is frequently had to these in such cases of obstinate hæmorrhage as are not attended by strong arterial action.

Pulmonary complaints, as obstinate coughs and asthmatic affections, have been said to give way to medicines of this class; yet, in modern practice, recourse is rarely had to them in such cases, and their exhibition is even considered as hazardous.

The ancients were accustomed to medicate some of their wines with resinous substances, the astringent flavour of which was agreeable also to their palates.[*] These wines were supposed to assist digestion, restrain ulcerous and other morbid discharges, provoke urine, and strengthen the bowels; but Dioscorides informs us, that they were known to produce vertigo, pain of the head, and many mischiefs not incident to the same quantity of vinous liquor free from such admixtures.

The particular preparations of turpentine most employed in the treatment of the several diseases alluded to will be noticed under the correspondent heads, which we have arranged in the botanical order of the species; and of which, therefore, the first is

SCOTCH FIR.

1. LIQUID RESIN.

(Resina liquida pinea.)
Terebinthina vulgaris. Dale's *Pharm.* (Ed. 3.) p. 278. Linn. *Mat. Med.* p. 153. *Pharm. Lond.* Common Turpentine.

THE *Common Turpentine* is more coarse and dense than any other sort, and has an opake light brown colour. Its consistence may be compared to that of honey. The taste is very acrid, hot, and disagreeable, and the smell much less pleasant than either the Venice or the Strasburg turpentine. It is this kind, which (as its name implies) is most commonly used on all occasions when a terebinthinate juice is wanted, either in medicine or in the arts, and a greater variety of preparations owe their origin to it than to the product of any other species of *Pinus.*

The artificial extraction of the resinous juice of the Pine seems to have been practised by the ancients in a manner very similar to that which obtains at present. Theophrastus [*] gives a particular account of the

[*] Kaauw de *Perep.* n. 450.
[*] See Dioscorides, lib. 5. c. 38, 36, 37, 38, 39, where he describes the Oivoç Πιτvιτης, Στροββιτης, Σαpιvoς, Ρητιvης, &c.
[*] Lib. 9. cap. 2.

several trees employed for this purpose in his time, of the proper season of the year for commencing the process, and of the several variations in the quality of the juice; and, though commentators have not been able to refer all the trees described by this venerable author to their proper places in the Linnean *Species Plantarum*, it is sufficiently evident that three or four kinds of turpentine were in use among the ancient physicians, which corresponded, in their properties, with those found in the shops of the moderns.' The mode of extracting the liquid resin of the Pine, however, is not mentioned by Theophrastus, or by any other writer of antiquity sufficiently in detail to deserve insertion here, and we shall therefore transcribe the account given by Duhamel, (in his "*Traité des Arbres*,") who is more circumstantial than any other author, and who, though not much more precise than the older naturalists in defining the particular species of trees most eligible for the operations, seems to have acquired very accurate information respecting the operations themselves, and several interesting circumstances connected with them.

It is well known that all Pines, even of the same species, do not yield an equal quantity of resin. Some produce three pints in one summer, and others not half a gallon the whole time they last. This difference does not seem to depend on the size or on the age of the tree, or, altogether, on the nature of the soil, because it is observable even in the same forest; but, in general, it has been remarked, that trees with the thickest bark, and which have been most exposed to the heat of the sun, yield the most. It is usual to select such as are of about four or five feet in circumference. At the foot of the tree a hole is made in the ground to the depth of eight or nine inches, and which will hold nearly a quart of the juice. Owing to the looseness of the soil in newly made pits, a portion of the juice of course is lost by transudation, but in mixing with the earth, it at length forms a mass sufficiently compact to resist any further draining. Though much attention is generally paid to cleaning the soil contiguous to the pits, sand, leaves, and fragments of bark will inevitably collect in the latter, and render a filtering process afterwards necessary. In some countries, a hole is cut in the substance of the tree itself near the root, in order to save the juice more free from impurities, but this practice is attended with danger to the former. When the proper receptacles are prepared, and a little while before the season for making incisions, the coarse back is stripped off, down to the *liber*, to the extent of about six inches. This precaution is the more necessary, in order that the edge of the instruments employed for making the incisions may not be injured; for if any splinters or filaments should be left in the wounds, the free course of the juice to the pits would be impeded; besides, in taking off the outer bark, it is scarcely possible to prevent fragments from falling down, and mixing with the juice, if any should have been collected, in the pits. As the resin flows most abundantly in hot weather, the incisions are begun near the end of May, and the extension of them continued to September. After the outer bark has been taken away, the inner bark and a thin slip of wood are cut off, with a very sharp tool, so that there may be a wound in the tree not more than three inches square by an inch deep; this first incision is made near the foot of the tree. Immediately after the operation, the resinous juice begins to exude, in very transparent drops, from the ligneous part and from between this and the bark; the bark itself yields scarcely any. The hotter the weather the greater is the production of resin, and the flowing ceases altogether at the approach of the chilliness of September. To facilitate the supply, the incisions are renewed once in three or four days, or oftener; for this purpose, the wound is a little enlarged, and a very thin slip taken off each time, so that an incision which, at the beginning of the summer, was only three or four inches in diameter, becomes, by the end of September, a foot and a half wide, and two or three inches deep. The following year a new wound is made just above the former, and managed in a similar manner. Thus, pines which have been cut for twelve or fifteen years have, one above another, twelve or fifteen wounds, and these several wounds reach to the height of as many feet, whence it becomes necessary to ascend steps to make the later incisions. It is not of much consequence on what side of the tree the incisions be made. The operators are commonly guided by the shape of the trunk, the situation of the ground, and the facility of digging pits; but there certainly seems to be some advantage in preferring that side of the tree which is most exposed to the sun. When the

---

* " Διαφορα μεν γαρ η τερμινθος, και γαρ πολυδενδρα, και ομοδενδρα, και καρποτε τε οφρα, αλλ' αλιγα, ωσπερ δε ς ελατος και στρωμα, κουφοτερα γαρ της πιτυινης. πλειστε δε " η ποιουσι και βαρυτατα και πυλωδεστερα, δια το μηδεποτε εκδεδεναι του πεποτι. (Ibid.)

By comparing the above description of the *juices* of terebinthiniferous trees with that given by modern writers on the *Materia Medica*, we should almost be led to decide, with positiveness, that the kinds enumerated by Theophrastus were no other than the Cyprian, the Strasburg, and the Common Turpentines of the European *pharmacopœia*: at any rate, from the description of the (*πιτυς*) *πινος*, we cannot but conclude that Bellonius has erred egregiously in considering the *ελατη* of Theophrastus as the *Picea*, and the *πευκη* as the *Pinus* of the Roman writers, a conclusion confirmed by a passage in Scribonius Largus, who, in speaking of the *resina pityina*, defines it in these words, " *id est ex picea arbore*." (See *Comp. Med.* c. 88.)

s M

pits are filled to a certain height, the juice is taken out with ladles made either of wood or iron, and poured into pails, in order to be removed to the hollow trunk of a Pine sufficient to hold three or four barrels.

The timber of Pines that have yielded resin even for fifteen or sixteen years is not the less valuable for domestic purposes; and it is a common opinion among the makers of tar, that the roots of such trees produce a greater quantity of that substance than those which have never undergone incisions.

Theophrastus mentions a disease to which Pines are subject, and which seems to consist in a redundance of resin collecting about the roots.[c] This disease, he says, proves fatal to the tree.

In France, distinct appellations have been given to the several states of the resinous juice of Pines, that which condenses on the wounds towards the decline of the sap being called *Galipot*, in Provence, and *Barras*, in Guienne; the fluid resin obtains the name of *Perinne vierge*; and a thinner kind of the latter, subjected to a sort of filtration, is called *Bijon*, or *Terebinthine fine*. The Galipot is used by the chandlers in making flambeaux, though the greatest quantity of this substance usually undergoes conversion, by being boiled, into *Brai-sec* and *Yellow resin*, to be described presently.

The liquid resin of the Pine, though of inferior quality to that of the Turpentine-tree, the Larch, and the Silver Fir, especially for internal use, is too often substituted for the others in the shops of the druggists. In most terebinthinate preparations this species is the subject, and there is no reason, perhaps, why the essential oil and other parts of it, separately taken, should not be equally good, and possessed of the same properties as what might be extracted from the juice of different trees. Whilst mentioning the essential oil, it may not be amiss to remark, that this seems to be the most active principle contained in turpentines, the several preparations of these juices manifesting most efficacy according as they are most impregnated with it; hence, in most cases, the common *Oleum Terebinthinæ* seems preferable to the crude resin; but of this oil we shall speak more particularly hereafter. The Colleges of London and Edinburgh direct the common turpentine to be used chiefly in external applications, for which it was much employed by the ancients also. Celsus mentions " *Resina liquida pinea*," as entering into the composition of many of his *malagmata*, and the " *resina liquida*" of other writers would appear to be of the same kind. The *Emplastrum Lythargyri compositum*, and *Unguentum Elemi compositum*, (Ph. Lond.) both contain this resinous juice as a principal ingredient; its digesting, cleansing, and incarnating properties, are acknowledged by medical practitioners universally. But its use is not confined to the healing art. In common life it helps to form many materials of no small utility, the poorer ranks of people, in many countries, making candles with it, the masons employing it in some of their mastics, the tinners in soldering, and the tallow-chandlers (when it is amalgamated with suet and yellow wax) for making flambeaux.

EXTRACT OF THE JUICE.

(*Extractum resinosum Pineum.*)

The pure resinous extract of the juice of the Pine has been recommended by some foreign writers for the cure of *gonorrhœa*, and is supposed to possess properties similar to those of the Peruvian and Copaiva balsams. I am not authorised, however, by any personal observation, or by accounts from any of my brethren in England, to mention this preparation in terms of commendation.

YELLOW RESIN.

Terebinthina cocta. Pharm. Wurt.
Resine jaune, of the French.

The mode of preparing this substance is minutely described by the French author whose name we have mentioned above.[d] He informs us, that the resinous juice is put into a large copper placed over a furnace, which

---

[c] This he calls "σπληϰ γασθεν," (Lib. 3. c. 10.) The word δας seems to have been misunderstood by Pliny, who speaks of the *Teda* as being a tree itself, instead of a disease incident to it, (see Lib. 16. c. 18.) which those who will take the trouble of examining the passages in Theophrastus will plainly see could not have been the meaning of the Greek naturalist.

[d] Duhamel. tom. 2. p. 145.

last is usually constructed with a mixture of clay, sand, and straw. Great care is taken that the sides of the furnace should adjoin close to the copper, lest the smoke of the fuel should mix with that of the resinous juice; for, without such a precaution, the heat of the furnace would not fail to set fire to the latter, and there would be a great risque of losing the whole; as an additional safeguard, there is generally a vaulted canal, four or five feet long, affixed to the mouths of the furnace, and terminated by a thick mud wall five or six feet in height. When every thing has been thus prepared, a moderate fire is kept up with very dry wood, and the juice boiled five or six hours, the operators frequently stirring it about with a large wooden ladle to prevent the impurities which sink to the bottom of the copper from taking fire, a circumstance likely to occur, it is said, without such a precaution. To ascertain whether the resinous matter is sufficiently boiled, a small quantity of it is taken out of the copper, and poured on a piece of wood; if, when it becomes cold, it may be reduced to powder by being pressed between the fingers, they know that the process is complete, and then conduct it out of the copper into a large trough similar to what is used for receiving the raw juice from the pits, and placed in like manner on supports. It is necessary to filter the decoction, which is done by pouring it hot on some long straw neatly stretched over wooden bars which form a kind of horizontal grating; the thickness of the straw filter is generally four or five inches. The impurities and dregs remain on the filter, and the juice runs through it pure into the trough. Before the juice becomes cold and fixed, it is let through a hole in the bottom of the trough into barrels, where it is suffered to harden; in this state it assumes a brown colour and a brittle texture, and is called *Brai-sec*, or *Rase*. To convert it into yellow resin: instead of turning the hot juice at once into barrels, an eighth part of fresh water is mixed with it in the trough. The water is acted upon so briskly by the hot decoction, that the whole continues to boil an hour or two, and the resin, from a brown colour, becomes at length of a fine yellow. It is afterwards deposited in barrels and suffered to harden like the *Brai-sec*.

The yellow resin, as ordered by the London College, is the result of a different operation, but the properties of both these substances are so extremely alike, that they may be spoken of under the same head; we shall therefore reserve our account of them until we have described the mode of preparing the

ESSENTIAL OIL.

(Oleum essentiale Pineum.)
Oleum Terebinthinæ. Pharm. Lond. & Ed.
Spiritus Terebinthinæ. Pharm. Wurt. &c.
Esprit de Rase, of the French.
Essential oil of Turpentine.

The process for obtaining this oil, as directed in the London Pharmacopeia, consists in distilling five pounds of the resinous juice with four pints of water, in a copper alembic. If one pound of the oil be redistilled with four pints of water, the result is called *rectified oil of Turpentine, (Oleum Terebinthinæ rectificatum,* of the London and Edinburgh colleges.) The process of rectification is not unattended with danger, for, unless the luting be very close, some of the vapour is apt to escape; and, if the latter should take fire, the vessels will unavoidably burst. In some dispensatories, this rectified oil is denominated *æthereal.* It does not differ very considerably in specific gravity, smell, taste, or medical qualities from the common essential oil.'

The *Oleum Terebinthinæ,* as we have before remarked, seems to be by far the most active part of the liquid resin of the Pine, and is, on that account, much more frequently employed in medicine than any other preparation. Its exhibition, however, requires considerable caution, and the admonitions of Boerhaave, Lange, &c. on this subject, cannot be too strictly attended to. The former of these authors, though its panegyrist, speaks of its violently affecting the head, producing bloody urine, and dangerously irritating the whole habit, when given injudiciously or in too large doses; and the observation of practitioners in general tends to confirm this assertion. Hence, it is proper to employ a very few drops at first, and not to augment the dose without great circumspection. The best vehicle of this powerful medicine is honey, which, with

' Edinb. Dispensatory, (1797) p. 315.

2 N

the admixture of a due proportion of powdered liquorice root, forms a good electuary.[a] Its use in diseases of the kidneys originating from ulcerations and obstructions[b] in those organs, is unquestionable; which is perhaps more than can be said of its alledged virtues in other complaints, and of those there are many for which medical writers of different ages have extolled the advantages of exhibiting turpentine. Cheyne recommends it as a perfect cure for *sciatica*; but, if I may be allowed to offer the result of my own professional experience, its effects are in few instances successful for the removal of that tormenting disease; and even those cases which I have seen cured, under the use of oil of turpentine, appeared to be rather of the symptomatic than of the idiopathic kind. It is reasonable to presume that the sciatic nerve, from its origin and course, may owe some of its morbid affections to an obstructed ureter, as well as to a rheumatic diathesis. In watching the state of the urinary excretion after the exhibition of turpentine, in more than one case of what is commonly called *sciatica*, I have actually witnessed considerable changes produced in it, and ascertained the pain about the hip to be mitigated according to the increased presumption of altered action in the ureter. The efficacy of *Oleum Terebinthinæ* as a *styptic* has been spoken of by some practitioners, but I have not myself witnessed any decided advantages produced by it, and, from having much more reason to confide in other medicines of that class, of late I have ceased to employ it; though, in uterine discharges attending cold, enfeebled habits, the more stimulative preparations of turpentine may certainly be exhibited with more safety than in the generality of diseases for which they are said to be calculated. As a diaphoretic, in rheumatic and gouty complaints, there are not wanting authorities for the employment of this medicine, but, in modern practice, it is rarely resorted to. Neither have the solvent effects which it has been said to produce (and which seem to have been inferred only from what is known to take place *out* of the body) on biliary *calculi* received much attention in the present day. In Germany, Norway, and some parts of the Russian empire, the essential oil of the Pine is frequently used as a remedy for lesions of tendons, and for bruises in general. In England, this remedy has repute principally among farriers, but the recommendations of authors so distinguished as Heister, Platner, and Plenck, certainly entitle it to more frequent trial in chirurgical cases.

But the use of the oil of turpentine is not confined to medicine. It is much employed by the painters for rendering their colours more fluid; and the concrete resins are usually dissolved in it when they are to be converted into varnishes.

COMMON RESIN.

(Resina arida Pinea.)
Resina flava. Pharm. Lond.

Is the residuum of the process for obtaining the essential oil. This process pushed as far as the nature of the substance will admit of changes the colour to a deep brown or black, when the resin acquires the name of

BLACK RESIN, or COLOPHONY.[c]

Resina nigra.

The medicinal properties of these two kinds of resin are, of course, extremely similar. They are rarely used internally, but for external purposes (particularly as plasters) they can scarcely be dispensed with,

---

[a] The best *formula* for preparing such an electuary is given in the pharmacopœia of Stockholm, which directs half an ounce of the oil to be mixed with one ounce of the best honey and as much liquorice powder as will make the whole of a proper consistence.

[b] I remember a remarkable case of hydatids formed in the kidneys, which came under my care in the Westminster Hospital, and which was very materially relieved by *Oleum Terebinthinæ* given in the dose of about eight drops every four hours. The expulsion of the hydatids seemed to be owing principally to the medicine, for, if the latter was omitted for a few days, the pain of the loins, *dysuria*, and general distress increased; and on resuming it, these symptoms were immediately alleviated whilst the hydatids were voided in augmented numbers.

[c] Essay on the Gout, (ed. 10) p. 119.

[d] I know not how this word came to be applied to the hard resin artificially extracted. It was originally the appellation of a raw liquid resin brought from Colophon, in Ionia, which is described by Dioscorides, (Lib. i. c. 77.) Galen *(de Comp. Med. Lib. 7.)* and Pliny, (Lib. 14. c. 20.) Celsus allows the choice of either the *Resina Colophonia,* or the *Resina Picea,* in the composition of his *sbentient* plasters. Scribonius mentions Colophony as a purgative, (c. 137.)

being remarkable for their adhesiveness, especially when mixed with other materials. Being deprived of the essential oil, these resins do not produce the same stimulating effects as other preparations, and may be considered as possessing astringency without pungency.

Colophony is of considerable use in the arts. It enters into the composition of several varnishes, and is sometimes substituted for sandarach. Musicians rub the bows and strings of violins with it, in order to take off the more greasy particles, as well as to counteract humidity.

TAR.

(Pix liquida Pinea.)

Πίσσα, Πίσσα υγρα, Κωνος of the Greeks.
Pix liquida, of the Romans, and of most modern *Pharmacopœiœ*.
Goudron, of the French.

This well known substance,' obtained from the roots and other parts of old Pines by a sort of *distillatio per descensum*, differs from the native resinous juice in having acquired a disagreeable empyreumatic quality from the action of the fire, and in containing the saline and mucilaginous parts of the tree mixed with the extractive and the oily. The Scotch Pine is the species from which most of the Tar used in this country is procured, and perhaps yields it equally good with its congeners. It is curious to remark how little the process employed in many countries differs from that which was followed by the ancient Macedonians, and which is circumstantially described by Theophrastus, in the third chapter of his ninth book, where he tells us that the billets were placed erect beside one another, and that they were afterwards covered with turf to prevent the flame from bursting forth, in which case the tar was lost. The stacks were sometimes, he says, one hundred and eighty cubits in circumference, and sixty, or even one hundred, in height.' These huge heaps of wood being set on fire, the tar was made to flow from them in channels cut for that purpose. As all the trees of this genus yield the same substance when treated in a similar way, it is probable that the ancients did not confine themselves to one species for obtaining it, any more than the moderns, and that some variety was occasioned in the product according to the different management of the fire, and in the cooling. Hence arises the confusion, and the difference of opinion among commentators respecting the terms *Cedria, Cedreleon, Pisscæleon*, &c. which, after the most industrious collation of passages from Theophrastus, Dioscorides, Galen, and Pliny, it is scarcely possible at this day to refer to the precise substances which they were intended to designate. But, we shall now proceed to point out the mode of procuring tar which Duhamel' states to have been practised in the Valais, and which seems to be the best that has been adopted.'

It is usual to cut down the Pines intended to be burnt for the extraction of Tar, in the course of the summer. The operators, knowing the quantity that will be wanted, regulate the extent of the hewing and tearing up of the trees so as that the materials may be neither too green nor too dry at the time of preparing their ovens, for, to make good tar, they should not be more than half dry. As all parts of the Pine (the trunk, branches, and even the bark) yield this substance, the branches are cut of a length proportionate to the size of the oven, and the thicker parts chopped into little billets similar to what are used in faggots. The ovens are shaped like an egg placed on its smaller end, and are composed of earth and stone, the floor being formed of one or more pieces of freestone, which are very nicely joined and hollowed like the inside of an egg-shell. On one side there is a hole about an inch and a half in diameter, and six inches in depth; to the external orifice of this, and five or six inches higher than the bottom of the oven, a gun barrel of a large caliber is affixed, and there is a large iron grate placed at the bottom of the oven. The dimensions of the oven vary according to the quantity of wood intended to be burnt, the largest being about ten feet high,

---

* As a summary definition of Tar, the words of Pliny perhaps cannot be improved. " *Pix nihil aliud est*, (says he), *quam combusta resinæ fluxus.*" (See Lib. 22. cap. 1.)

' " Ει εγκεται κουλον κ εγκλανεολα και εκολον επχραμι, πε οξεσι δι εξανενω, πελτοι γ αιπγεαικεα, η πεντεο αμφοριας." (*Loc. supra citat.*)

' *Traité des Arbres*, Tom. 2. p. 160.

' The reader may be gratified in consulting the work of John Conrad Axtius, (entitled, " *Tractatus de Arboribus Coniferis et pice conficienda.*" Iona. 1679. 12°) which contains much curious matter on this subject, as well as on the products of Pines in general.

2 O

five or six feet in diameter in the middle, and two feet and a half at the mouth or superior part. The walls are about a foot and a half in thickness. To about two-thirds of the height these are constructed with free-stone, but above that with oven-earth. When the ovens are finished, and quite dry and tight, bundles or faggots of the wood, tied up with hazel or vine rind, are set upright on the grating; the ligature is cut by means of a blade fixed at the end of a stick; and the pieces are spread about, the interstices being filled with chips. This first layer being properly made, a second faggot is let down, then a third, and so on until the oven is full, as high as the hand can reach, when chips and shavings are laid on, to the thickness of three or four inches, and the mouth is covered up with flat stones piled one upon another so as to close all gaps except at the center, where an opening is left four or five inches in diameter. All things being thus prepared, the chips at the top are set on fire, and the operators, who from experience are enabled to ascertain when the materials are sufficiently kindled, seize the proper time to shut up the mouth entirely with a flat stone; and they stop up with earth every interstice from which smoke is seen to escape. The wood then becomes reduced to charcoal, and the resinous part of it, mixed with the sap, flows through the grate down into the cavity at the bottom of the oven. When this cavity is full up to the place where the iron tube is fixed the tar flows into barrels placed to receive it. It is from custom alone that the persons who superintend the operation ascertain when the wood has given out all its resinous liquor; they then open the top of the oven, removing the stones, and collecting the soot which lodges in their interstices as well as on the sides of the oven, and which forms a kind of *lamp-black*; lastly, they take out the charcoal that has lodged on the grating, and recommence the operation by laying on wood as before. Such impurities as are heavier than the tar, with which they were mixed, remain on the stone that serves as a floor to the oven, whilst the tar itself flows on the surface through the tube, which, as we have remarked before, is five or six inches above the level of the stone. As far as we can judge, all the art of the operation consists in a proper management of the fire, for, if the oven be too closely stopped, the fire is extinguished, the wood is but imperfectly charred, and very little tar is extracted; but if, on the contrary, the wood burn too briskly, a great proportion of the resinous matter is consumed. When the fire is properly regulated, there is no flame in the oven; the heat and smoke, which are reverberated on the wood, cause the resin and sap to flow from the latter together. It would seem that a more certain mode of regulating the heat would be, instead of closing the top of the oven with stones and turf, to adapt registers of different sizes to a kind of dome, which might form the upper part of it, and render the structure more neat and commodious.

Tar has been used as a medicine both externally and internally. The ancients had a high opinion of its efficacy in pulmonary diseases, supposing it to promote expectoration, relieve dyspnœa, and check spitting of blood. Dioscorides particularly speaks of its utility in these cases. He also recommends it to be applied to ulcers, which, he says, it fills up and heals, whether they be situated on the surface of the body, or in the ears, throat, or other internal parts. In fact, there is no end to the praises bestowed by medical writers on the properties of tar, which, if we are to give credit to all the accounts given of it, is equal to the cure of all the maladies of the human frame. The colleges of London and Edinburgh direct it to be made into an ointment (*Unguentum Picis*); the former, by means of the admixture of an equal portion of mutton-suet, and the latter, of two-fifths of yellow wax. This ointment has been employed for the cure of cutaneous affections, particularly those of domestic animals. Some practitioners have applied a plaster of tar for the cure of obstinate cases of *Tinea capitis*, and not without success; but it is a very painful, and almost a cruel, remedy, for it cannot be taken off without dragging out of the skin adhering to it the roots of the hair, in the eradication of which, in fact, consists the only use of the plaster.

TAR-WATER.

(*Infusum Picis liquidæ pineæ.*)
Aqua picea. Pharm. paup.

WATER impregnated with the more soluble parts of tar, and hence called *Tar-Water*, was once a very popular remedy for various obstinate complaints, both acute and chronic. It was indebted for its great reputation principally to the celebrated Dr. George Berkeley, Bishop of Cloyne, who wrote a long disser-

See also Pliny, lib. 24. c. 7.

tation on it, under the title of " *Siris, or a chain of philosophical reflexions and enquiries concerning the virtues of Tar Water.*" A narrative of its success, with a great number of cases and remarks, was published also by Thomas Prior, Esq. From the accounts given by these writers and by Cullen, (who appears to have entertained no mean opinion of its efficacy) it appears to strengthen the tone of the stomach, to excite appetite, promote digestion, and remove many dyspeptic symptoms, at the same time increasing the excretions, particularly that of urine. Cullen believed the virtues of this medicine to depend chiefly on the acid principle it contains, and it was on this account that the Bishop of Cloyne preferred the Norway Tar (made from *Pinus Abies*) to that of New England; this acid, however, does not appear to differ from that which is extricated by fire from all kinds of recent wood.' The proportions recommended by the Bishop are a gallon of cold water to a quart of tar; these are to be mixed thoroughly by means of a ladle or flat stick, with which the liquor is to be stirred for the space of three or four minutes; and the vessel containing it must stand forty-eight hours, that the tar may have time to subside; the former is then strained. Tar water distilled yields a liquor much impregnated with its flavour, though more grateful than the infusion itself both in smell and taste. There remains a light, spongy, blackish matter, not acid, but bitter, partially soluble again in water. Whatever commendation tar-water may formerly have had, or may justly be entitled to, it does not receive much in the present day, being but little used, on account of other medicines appearing to possess the same properties in a more exalted degree; and indeed there are other *terebinthinate* preparations better suited than this to answer the purposes for which it was formerly recommended, as will be mentioned in the course of this dissertation.

PITCH.

(Pix pinea inspissata)
ποπιας, of the Greeks. *Spissa Pix.* Pliny.
Brai-gras, of the French.

The usual mode of making pitch consists in melting coarse hard resin (or *brai-sec*, as it is called in France) with an equal quantity of tar, in large copper vessels similar to those used for boiling the raw juice. If the tar be too thin, the proportion of resin is increased; and, on the other hand, if it be thick, a third part of tar is sufficient. Should the process of inspissation be carried to its utmost limit, the pitch becomes quite hard and dry, and is called in the shops *Pix arida* (the ξηρα ρητα and πισσαωτα of the Greek writers), which is less pungent and less bitter than the common tar, and is used only in some external applications, as an adhesive substance agreeing in its medicinal virtues with the common digestives.

When melted with oils, resins, and fats into ointments and plasters, pitch is said to be very apt to separate and precipitate. Dioscorides describes the best pitch as being shining, odorous, gummy, and of a reddish black colour, which were the qualities of the Lycian and Calabrian pitch.' It was prescribed by him, and also by Celsus, as a proper ingredient in plasters for maturing abscesses and healing wounds.

Pitch was much employed by the ancients for giving flavour and fragrance to their wines, which were also supposed to acquire from it useful medicinal properties, as we have before remarked, when speaking of the properties of terebinthinate substances in general.' Their mode of pitching casks and other vessels is described by Columella.' We are told by Pliny of a preparation of tar with vinegar, called *Brutia*, which was employed for the same purposes;' and this author says that it was usual to sprinkle the first ferment of new wine, or *mustum*, with powdered resin.

In boiling down tar to dryness without addition, there comes over an acid liquor in considerable quantity, and also an ethereal oil, which seems to differ from the oil of turpentine only in being impregnated with an empyreumatic quality; it was called by the ancients *Oleum Picinum.* The medicinal properties of this oil are similar to those of tar.'

The extensive use of pitch and tar in ship-building is too well known to require particular mention. A mixture of pitch and wax, by which crevices in vessels are rendered impermeable to water, was called by the

' Lewis's Mat. Med. Art. *Pix liquida*.   ' Lib. 1. c. 81.   " See page 2.   ' " De re rustica." Lib. 12. c. 18.
        ' Lib. 16. cap. 11.   ' Dioscorid. Lib. 1. c. 79.
2 P

ancients *Zmærx;* and this substance, after it had been some time steeped in the sea, was used medicinally as a resolvent.' Blended with a certain quantity of oil and suet, pitch becomes an useful article to the shoemakers for waxing their threads, and with whale fat it forms the grease with which wheels of carriages are smeared over. In several kinds of luting also, this article possesses considerable utility, and is familiar to most mechanics and handicraftsmen.

LAMP-BLACK.

(Fuligo Pinea.)
Noir de fumée, of the French.

ANY species of Pine may be used for making lamp-black, but the general practice is to convert the impurities left in the precipitation of tar and pitch to this purpose. The mode followed in Germany is thus described by Axtius, who has been copied by Duhamel, and it is illustrated in the works of both these authors by engravings.' A sort of box is made, nicely closed in every part with the exception of some holes in the top, which are covered, however, with a sort of linen cone. At a little distance from the box a furnace is constructed, with a very small mouth, and the inferior part communicating with the inside of the box by an horizontal chimney. Into this furnace are put the dregs and coarser parts left in the preparation of tar, and in proportion to the consumption of these a supply is kept up, so as to furnish a constant draught of smoke to the box. The smoke goes chiefly into the cone, where it deposits its grosser parts in the form of soot, which when beaten off from the linen by sticks applied on the outside, is collected from the upper part of the box and put into barrels.

Lamp-black is employed almost exclusively in printing and dying in the present day, but it was formerly used as a substitute for the *Fuligo Thuris,* which is mentioned by Dioscorides and Celsus as a resolvent and digestive, and formed an ingredient in some of their plasters. The first of these authors describes a process for obtaining lamp-black literally by means of a lamp, and attributes to it astringent properties (especially in ichorous discharges from the eyes) as well as a remarkable efficacy in promoting the growth of hair on the eye-brows.' Galen also adverts to the same remedy in his account of the fuliginous substances prepared from different kinds of resin.' There is a *Tinctura Fuliginis* retained in the Edinburgh *Pharmacopæia;* this is exhibited internally, as an antihysteric, but rarely trusted to alone, being found most efficacious when combined with assafœtida, or other medicines of that class, to all of which it seems to be far inferior. It is directed to be prepared from *wood-soot,* without any particular tree being specified as preferable for this purpose to another.

BARK-BREAD.

WE are informed by Linnæus' that the Laplanders eat, during a great part of the winter, and sometimes even during the whole year, a preparation of the inner bark of the pine, which is called among these people *Bark-broed.* This substance is made in the following manner, viz. After a selection of the tallest and least ramose trees, (for the dwarf, branching ones contain too great a quantity of resinous juice) the dry and scaly external bark is carefully taken off, and the soft, white, fibrous, and succulent matter collected and dried. The time of the year chosen for this process is when the *alburnum* is soft and spontaneously separates from the wood by very gentle pulling, otherwise too much labour would be required. When the natives are about to convert it to use, it is slowly baked on the coals, and being thus rendered more porous and hard is then ground into powder, which is kneaded with water into cakes and baked in an oven.

The Siberian ermine-hunters, when their ferment or yeast which they carry with them to make their *Quass,* is spoiled by the cold, digest the inner bark of the pine with water over the fire during an hour, mix it with their rye-meal, bury the dough in the snow, and after twelve hours find the ferment ready prepared in the subsiding fæces.'

' Dioscorid. c. 82.   ' *Tract. de Arb. conif.* c. 15. *Traité des Arb.* Tom. 1. p. 17. fig. 4.   ' *Dioscorides,* lib. i. c. 79.
' *Simpl. Med.* lib. 7.   ' *Fl. Lapp.* Smith. p. 284.   ' *Pallas. Fl. Ross.* 1. p. 2. 3.

THE MUGHO PINE.

LIQUID RESIN.

(Resina liquida Pini Pumilionis).
Balsamum Hungaricum, of the German *Pharmacopœiæ.*
Hungarian Balsam.

This resin spontaneously exudes from the extremities of the branches, and from other parts of the tree, and may also be obtained, by expression, from the green cones. Its reputation as a medicine originated from a manuscript account written by Dr. Christian ab Hortis, of Käsmark, who extolled its efficacy in the cure of wounds, running ulcers, contusions, rheumatisms, palsies, and even of the gout.[a] Various other complaints were said to be cured by it, and it afterwards received the commendations of Fischer,[b] Breynius,[c] and Bruckmann,[d] the first of whom considered it not inferior to the Balsam of Mecca. In Germany, this balsam still retains high repute, but there can be no doubt that its medicinal virtues have been much exaggerated.

ESSENTIAL OIL.

(Oleum essentiale Pini Pumilionis).
Oleum templinum, of the German *Pharmacopœiæ.*
Krumholz-oil.

This essential oil is obtained by distillation from the resinous juice just described. The common oil of turpentine is often substituted for it by the itinerant druggists in Germany, but the genuine sort may be distinguished by its golden colour, agreeable odour, and acrid oiliness of taste.[e]

As a medicine, this oil is a popular remedy at Brunswick for the cure of intermittents, being taken, in the dose of a few drops, just at the commencement of the cold stage. It is also used in punctures of tendons, and by farriers as an application to foul ulcers of cattle.[f]

THE STONE PINE.

KERNELS.

(Nuclei Pineoli.)
πιτυϊς Diosc. Lib. 1. c. 74.
Pignons, of the French.

These kernels have a subacid, sweet taste, similar to that of almonds, and, like the latter, may be used for emulsions as well as for dissolving resins. They possess a nutritive and demulcent quality, but, from their oily nature, soon become rancid and unfit to be eaten. Dioscorides speaks of their utility in coughs, and it is probable enough that they act as expectorants in some degree; in the present day, however, they are rarely used, except at the table.[g]

The Siberian Stone Pine (*P. Cembra*) yields nuts of the same kind as these, which are, therefore, applicable to the same purposes, but their oily contents, when exposed to the air, manifest a still stronger disposition to acquire rancidity.

The proportion of oil in the kernels of these nuts is larger, perhaps, than in those of any other tree, one pound of them yielding five ounces, whereas the same quantity of linseed produces only two ounces and a

---

[a] See Murray's *App. Med.* Vol. I. p. 12. [b] Bresl. Samml. Vers. 2. p. 331. [c] Eph. N. C. Cent. 7. p. 5. [d] *Specim. dus.* 1787.
[g] *Hermann, Mat. Med.* T. 2. p. 194. [e] *Murray, Vol. I. p.* 13. [f] *Mat. Med.* p. 124.

2 Q

half.* Rhaze speaks of the oil having a tendency to relieve obstructed kidneys, a circumstance not improbable, and well deserving of being put to the test of experience, if it were only for the sake of substituting what would be so much more agreeable to the palate than the common turpentines.

THE SIBERIAN STONE PINE.

ESSENTIAL OIL.

(Oleum essentiale Pini Cembræ).
Balsamum Carpathicum, of the Germans.
Balsamum Libani. Murray. *App. Med.* Vol. I. p. 15.
Carpathian Balsam.

The German writers describe the Carpathian Balsam as being pellucid, very liquid, and of a whitish colour, and as having an odour and taste like oil of Juniper. We are informed that it is extracted, by distillation, from the shoots of *Pinus Cembra*, after they have been bruised, and macerated a month in water. It was brought into notice by the authors who are mentioned above as having treated of the Krumholz-oil, and they have recommended it to be rubbed on the temples, for the cure of *vertigo*, and to be dropped into the ears for the cure of *susurrus* and difficulty of hearing. Like all other popular medicines, this balsam has reputation, in Germany, for performing great wonders in many complaints of very opposite characters, as pleurisy, erysipelas, *calculus*, and putrid disorders; hence its alledged virtues are not to be too hastily credited.

THE SILVER FIR.

LIQUID RESIN.

(Resina liquida pytuina).
Terebinthina Argentoratensis, or Strasburg Turpentine. Dale's *Pharm.* (Ed. 1737) 277.

This resin is generally of a middle consistence between that of the *Terebinthus* and the *Lariæ*; more transparent and less tenacious than either; in colour yellowish brown; in smell more agreeable than any other turpentine, except the Cyprian; in taste the bitterest, yet least acrid.

Our own shops do not now retain any liquid resin under the name of Strasburg Turpentine, but it is sold in several parts of Germany, where its reputation still exists; and, according to some authors,' this sort is to be considered as equal, if not superior, to all the other substitutes for the genuine juice of *Pistachia Terebinthus*.

The greatest quantity is collected by the peasantry living in the vicinity of the Alps, who set out every year towards the month of August, provided with sharp-pointed hooks and a bottle suspended to the waist, for the purpose of puncturing the vesicles adhering to the bark of the trees. It is a curious sight, says Duhamel,' to observe the peasants climbing to the very top of the highest firs by means of the cramp-iron with which their shoes are armed, and which pierce the bark, their two legs and one arm grasping the trunk, whilst the other arm is employed with the hook. Some of them make use of a pointed bull's horn, which serves both as an instrument and a receptacle, and, as soon as it is filled, the resin is poured into the bottle; the bottle is afterwards emptied into a larger or into skins, which are used for conveying the resin to the most advantageous places of sale.

As it often happens that leaves, fragments of bark, moss, &c. fall into the horns, the resin is purified by filtration; for which purpose a piece of bark is stripped off from a fir to make a kind of funnel; at the

* Haller *Hist.* (Tom. 2. p. 317.) on the authority of Spielman.   * Murray, p. 15.   * Matthiol. p. 107. 334.   Haller, p. 314.
* *Traité des Arbres*, Tom. I. p. 9.

narrow end of which are put some shoots of the same tree, and then this funnel is filled with the resinous juice which has been collected; the juice gradually flows through, and the extraneous matter remains enveloped in the shoots. This is said to be the only preparation used previous to sale. Very little resin is collected by incisions in the bark of *Pinus Picea*, the great supply being from the vesicles, which may be distinguished from those often formed in the pores of *P. Abies* by the juice of the former remaining clear and transparent like mastic, whereas that of the latter, in thickening, becomes opake like frankincense. The vesicles found under the bark of *P. Picea* are sometimes round, and sometimes oval, but it is remarkable that in the latter case the longest diameter is always horizontal, and never perpendicular. In places where the trees appear to be rich and the soil substantial, liquid resin is collected twice in the year, namely, in spring as well as in autumn, but no tree produces vesicles more than once during the ascent of the sap; in poor soils indeed vesicles are produced only in the spring.

The trees begin to yield a moderate quantity of resin as soon as they are three inches in diameter, and continue to do so until they are increased to a foot; at this period, the wounds made in the bark form hard and horny scales; the woody part continuing to enlarge forces the bark, which is hard and incapable of extension, to crack open, and in proportion to the increase of size in the tree, this bark, which, when the latter was young, did not acquire greater thickness than a quarter of an inch, at length grows to an inch and a half, and then it yields no more vesicles.

Lewis says that the Strasburgh turpentine is procured both from the silver and from the red fir, by cutting out, successively, narrow strips of the bark, from as high as a man can reach to within two feet of the ground; but we are informed by Duhamel (from whose work the above account is principally taken) that this practice is confined to *Pinus Abies*.

The uses to which the liquid resin of the silver fir is applied are, for the most part, so similar to those mentioned under the head of *Resina liquida larigna*, that we shall not detain the reader with any account of them in this place.

THE BALM OF GILEAD FIR.

LIQUID RESIN.

(Resina liquida balsamca.)
Balsamum Canadense. Pharm. Lond. et Ed.
Baume blanc de Canada, of the French.
Canada balsam.

This is a transparent whitish juice, brought to this country from Canada, and apparently not very different in its qualities from the celebrated *Balm of Gilead*, so high in esteem among the eastern nations, and so strongly recommended in a variety of complaints. Hitherto, however, it has not been much employed in England, yet it is thought capable of answering all the purposes for which the Copaiva balsam is employed, and would therefore deserve a more general trial. It has an agreeable odour, and a strong, pungent taste.

THE NORWAY SPRUCE FIR.

CONCRETE RESIN.

(Resina concreta abiegna.)
Thus. Haller. St. Helv. 2. p. 313.
Poix, of the French.

This substance spontaneously exudes from the pores of the tree, and soon concretes into distinct drops, or tears, which differ from the produce of the Silver Fir in being compact, opake, and of a deeper yellow

---

* Mat. Med. (Ed. 4.) Vol. II. p. 417.
† The product of *Amyris Gileadensis*, and probably the *Balsamum Judaicum, Syriacum, è Mecca, Opobalsamum,* &c. of the older writers.
‡ The *Thus* of the ancients does not appear to have been the product of any species of *Pinus*, but, as we are informed by Dioscorides, (Lib. 1.

2 u

yellow colour. The common frankincense of the shops is probably no other than this resin, or, at least, the latter is in general mixed with the former, and becomes an ingredient in the *Emplastrum Thuris compositum*, and *Empl. Ladani compositum*, of the London *Pharmacopœia*, applications employed as corroborants and constringents, but useful chiefly from their mechanical properties, by which considerable support is afforded to the integuments whereon they are laid.

BOILED RESIN.

(Resina abiegna cocta).
Pix Burgundica, Pharm. Lond. &c.
Poix-grasse, of the French.
Burgundy Pitch.

THIS substance is of a close consistence, but rather soft, of a reddish brown colour, and not unpleasant smell. The shops are supplied with it chiefly from Saxony, where, however, many adulterations take place, and, in this country, it is to be suspected that a preparation of the common turpentine is sometimes substituted for it.

The genuine Burgundy Pitch is prepared from the resinous juice of the Norway Spruce, which is yielded by that tree from the month of April to September, from the effect of incisions made in the bark. In the operation of cutting, the wood is left untouched, for the juice exudes chiefly from *between* the bark and the wood, and in small quantities from the former alone, but not at all from the latter. It fixes almost immediately after it is freed from its lodgments, and therefore does not flow to the ground, but remains attached to the tree in large tears, or flakes. This concrete matter is collected once in a fortnight, by detaching it with an instrument formed on one side like an axe, and on the other like a gorget. With this instrument also the incision is renewed every time that the resin is collected. The resin of young Spruces is softer than that of old ones, but it is never fluid, except in hot countries, and under considerable exposure to the sun; and even in those circumstances it has not the same liquidity as that of the Pines properly so called. In forests impenetrable to the rays of the sun, the incisions are usually made on the south side, but in different situations this is not always the practice; the side most exposed to rain is never subjected to incisions. If only one incision be made in a spruce, there will be a supply of resin from twenty-five to thirty years. A vigorous tree, planted in a good soil, will yield in one year thirty or forty pounds of juice. This juice becomes dry enough to be put into sacks, and is thus conveyed by the peasants to places where the following process is carried on, viz. The resinous substance is put, with a proper quantity of water, into large boilers; a moderate fire melts it; it is then strained, under a press, through strong, close cloths, into barrels, to be transported to any distance. It is rarely cast into loaves, because these would be melted by a low degree of heat, and easily put out of shape.

The above is the mode practised in the neighbourhood of Neufchatel,[c] but it is different in other places, as will be seen by consulting Axtius,[d] Geoffroy,[e] &c.

Burgundy pitch is employed only externally. Formerly, it was an ingredient in a great variety of ointments and plasters, but at present, its use is confined by the London College to the *Emplastrum Cumini*, and to another which takes its name from this resin, viz. the *Empl. Picis Burgundicæ compositum*. It forms a warm, stimulant application, and sometimes creates even vesications, though, in general, a redness of the part, with a gentle exudation, is the only effect observable. The cases in which the *Empl. Picis Burg. comp.* seems to produce most good are those which may be called *nervous dyspnœa*; but it is serviceable likewise in coughs, pains of the muscular parts of the chest, and some affections of the trachea occasioning loss of voice. It should be renewed once in three or four days, and so continued for a fortnight or longer.

cap. 70.) pine resin was often substituted for it; and the substance now bearing that name in the shops is seldom any other than the concrete resin described above. Dioscorides describes a mode of distinguishing the two kinds. " Resin (of the Pine) says he, when thrown into the fire " dissipates itself in smoke, whereas frankincense burns with a brisk flame, and the odour of the latter serves to detect imposition." Some writers are of opinion that the genuine *λίβανος, Thus*, or Frankincense, is obtained from *Juniperus Lycia*, and constitutes the *Olibanum* of our shops, but I cannot find any passages in ancient authors sufficiently precise to corroborate this conjecture.

[c] Duhamel. Tom. i. p. 15.   [d] *Tract. de Arb. conif.* p. 79.   [e] *Mat. Med.* T. 3. p. 437.

BRANCHES.

(Ramusculi abiegni).

THE *effluvia* of the Norway Spruce are supposed to effect a salubrious impregnation and coolness of the air, on which account, it is usual in Sweden, to cut the branches into pieces of about half a finger's length, and strew them on the floors of apartments tenanted by invalids.' Jonston in his *Dendrographia*, speaks of the wholesomeness of walking in groves of Pines; and Linnæus informs us that the Laplanders apply the young shoots to the head, for the removal of pains in that part.'

THE WHITE SPRUCE FIR.

SPRUCE BEER.

(Decoctum Abietis compositum).

THIS drink seems to be the best of any that is made from parts of firs or pines, being not only most agreeable to the palate, but also most answerable to the indications of cure in those complaints for which decoctions and infusions of cones, tops, leaves, &c. of various species, have been prescribed. The *Decoctum Turionum Pini*, the *Essentia Abietis* (of the Augustan college) and other forms prescribed for the prevention and removal of scurvy, will probably soon give way entirely to the American Spruce beer, the mode of preparing which we shall extract from the valuable French writer, of whose work we have availed ourselves so much already.'

To make a cask of Spruce beer, there ought to be a boiler large enough to hold one-fourth more. This is to be filled with water, and as soon as the latter begins to boil, a bundle of Spruce branches broken into pieces is to be thrown into the boiler; the bundle should be about twenty-one inches round, at the place of ligature. The water is to be kept boiling until the rind becomes easily detachable from the branches, and whilst this process is going on, a bushel of oats are to be roasted several times over in a large iron pan, and fifteen sea-biscuits, or, instead of these, twelve or fifteen pounds of bread cut into slices, should be well browned, to be mixed altogether with the liquor in the boiler. The branches of Spruce are then to be taken out, and the fire extinguished. The oats and bread fall to the bottom. The leaves, &c. floating on the surface of the liquor being skimmed off, six pints of molasses, or coarse syrup of sugar, or, in default of these, twelve or fifteen pounds of brown sugar are to be added. This mixture should be immediately turned into a fresh port-wine cask, and, if it be intended to give a colour to the beer, the dregs, and from five to six pints of the wine, may be left in the cask. Whilst the liquor remains tepid, half a pint of yeast must be added, and briskly stirred about, in order to incorporate it well with the decoction, after which the cask is to be filled up to the bung-hole, and the latter left open. The liquor will ferment and throw off a good deal of excrementitious matter; in proportion to the quantity that works out, the cask is to be replenished with some of the same decoction kept apart for the purpose. If the bung-hole be stopped at the end of twenty-four hours, the spruce remains sharp like cyder; but if it be intended to drink it softer, the bung must not be put in until the fermentation is over, taking care to replenish the cask twice a day.

The writings of physicians and voyagers abound with testimonies of the antiscorbutic virtues of Spruce-beer, the reputation of which seems now to be so general and so well founded that it is wholly unnecessary to dwell on this subject. The beverage has not only its wholesomeness, but becomes pleasant to the palates of most persons who have habituated themselves to its use. From its refreshing and strengthening qualities, it is very generally had recourse to as a common drink during the heats of summer, and is considered

' Murray, Vol. I. p. 29.   ' *Flora Lapp.* (Smith's Edit.) p. 287.
' The reader may consult also the description given by Kalm, in the *Trans. of the Acad. of Sc. of Stockholm* for 1751, p. 190. This traveller's countryman, Dr. Arvid Faxe, published a mode of making this sort of drink from the Scotch Fir, which, as well as many other species of *Pinus*, may, no doubt, be made to answer where the Spruce firs are not to be found. Duhamel thinks that juniper may be substituted for spruce.

almost as much a luxury at the table as a valuable article of medicine. By what particular principle in the liquor its power of preventing and subduing scorbutic complaints is effected, or whether this efficacy be not the result of the general preparation, does not appear to be as yet determined; but, as simple decoctions and infusions of the branches of several species of *Pinus* have been found conducive to the same purposes, it is most reasonable to suppose that the spruce alone is the essential part of the medicine. Some authors have imagined the salutary properties of medicines of this sort to consist in their gently increasing the secretions from the kidnies; whilst others have contended for a peculiar power in the vegetable acids to promote venous absorption. Whatever be the *modus operandi*, however, the acid contained in infusions of fir, and which has been called *Acidum Abietis*, has been found, when exhibited by itself, to act as a diuretic,[*] and, though the properties of this acid may be no way different from those of others extracted from vegetable substances, it is highly probable that this is one of the most active principles in terebinthinate infusions.

THE LARCH.

LIQUID RESIN.

(Resina liquida Larigna). Largatum of the Italians.
Terebinthina Veneta. Pharm. Ed.
Venice Turpentine.

This resin is by most writers, and in the shops, esteemed the best (after that of *Pistachia Terebinthus)* of those juices commonly called *Turpentines.* It is usually thinner than any other sort, of a pale yellowish colour, and of a hot, pungent, bitterish taste. The smell is strong, and far from being agreeable. Though it bears the name of *Venice Turpentine,* very little of it is exported from the Venetian territories; but it is probable that the merchants of that country were the first who substituted it for the genuine Turpentine of Cyprus. That which is most commonly met with in the shops comes from New England; from what tree we are uncertain. The true liquid resin of the larch is obtained chiefly from France and Germany.

The extraction of the juice of the Larch is performed by boring holes, of about an inch in diameter, and with a gentle descent outwards, in the most knotty parts of the tree, proceeding from the height of two inches above the ground to ten or twelve. The south side is generally preferred, on account of the sun's heat facilitating the flow of the juice. There are adapted to the holes what may be called gutters, which are fifteen or twenty inches in length, and terminate like pegs perforated in the centre. The juice passing along these tubes falls into troughs placed at their most depending 'part. The result of the process is carefully examined every morning and evening, from the end of May (which being the time at which the trees are fullest of sap, is also that at which they are generally perforated) to the end of September. As soon as the supply appears to diminish, the holes to which the gutters were attached are stopped up with pegs for about a fortnight, after which interval they are opened again, to discharge the re-collected resin. This operation is repeated until the tree being drained furnishes but a very small quantity of juice; the pegs are then replaced, and not withdrawn until the succeeding season. As it is impossible to secure the troughs from leaves, and other extraneous matter which will mix itself with the fluid, the latter is usually strained through a hair sieve into other vessels, in order to be transported to the places of sale. It is only from the most healthy and vigorous trees that the resinous juice is obtained; those which are too young or too old produce only a small quantity, though at all ages tears of this substance may be seen on various parts of the trees, and especially where knots exist. Some of the mature trees (Haller says[ᵃ]) will discharge five pounds of juice in a year. The resin seems to reside in the exterior circles of the wood, for, if the healthiest part of the tree be cut into billets, there are often found, at the distance of five or six inches from the inmost part, *dépots* (as it were) of this resinous matter, which are sometimes an inch in thickness, three or four inches in width, and as many, or more, in length. In a tree thirty or forty years old, it very frequently happens that five or six of these

[*] The late Dr. Hope made frequent trial of this acid in the Royal Infirmary of Edinburgh, and conceived it to be beneficial in obstinate coughs and catarrhs. (See Lewis's *Mat. Med. Art. Abies.*)   [ᵃ] *Stirp. Indig. Helv.* Tom. 2. p. 314.

reservoirs are found, and a multitude of smaller ones, which occasion great inconvenience in sawing, on account of the teeth of the saw being clogged by the glutinous matter.[1]

The resinous juice of the Larch is said to remain always, or at least a very long time, in a state of liquidity; if it should at length become at all concrete, it would be only at the edges, or on the sides of the vessel in which it may be contained.[2] This property is adverted to by Pliny.[3]

The resinous juice of the Larch is substituted, in medicine, for that of the true Turpentine tree under those circumstances to which the latter is supposed to be particularly suited. As a diuretic the Venice turpentine is generally preferred to all the other sorts, and it loosens the belly most, on which account Riverius[4] thinks it more safe than most other irritating diuretics. Some authors have thought that it had a greater tendency to produce vertigo, and even drunkenness, than other terebinthinate juices, when taken internally, but this is not well ascertained. Cullen[5] observes that its effects, when it is employed in an *enema*, are more certain and durable than those of saline medicines, for remedying obstinate constipations of the bowels. Other practitioners have employed it with advantage in some nephritic cases, exhibiting it in a similar form, by the *rectum*. With regard to external use, this resin is much employed in the Tyrolese territories for the cure of wounds. It forms a part of several plasters and ointments, as, for instance, the *Emplastrum de Belladonna*, of the Brunswick pharmacopœia, the *Unguentum Infusi Cantharidum*, of Edinburgh, and the *Ung. digestivum*, of the Russian shops, being esteemed not only on account of its mechanical uses, but as an efficient digestive and discutient.

G U M.

(Gummi Larignum.)
Gummi Orenburgense, of the Russian shops.

This gum does not appear to be used any where but in Russia, yet it is described as a good substitute for *Gum Arabic*, of which it possesses much of the glutinous quality. It is dryish, of a reddish colour, and of a sub-resinous taste, but wholly soluble in water.

The mode in which this substance is obtained is very remarkable.[6] It sometimes happens that whole forests of Larch, in some parts of the Russian empire, are consumed by fire, in consequence of the flames being driven from the open hearths of the huntsmen to the foliage of the trees by a high wind. During the combustion of the medullary part of the trunks, a gum issues forth, which is diligently collected by the natives for the purpose not only of rendering their bows glutinous, but also of being eaten as a delicacy. It is likewise supposed to act medicinally, as an antiscorbutic, and an useful astringent of the gums.

MANNA OF THE LARCH.

(Manna Larigna.)
Manne de Briançon, of the French shops.

About the month of June, and when the sap of the Larch is most exuberant, this tree produces small white drops of a sweet, glutinous matter, like Calabrian manna. The young trees generally produce most, appearing quite white with it early in the morning before the rays of the sun have acted strongly and dissipated it; and, what is very singular, it adheres almost exclusively to the extremities of the branches. Bellonius[7] remarked that (as was the case with the Cedars of Mount Libanus) the drops are discoverable only on those trees which occupy the highest spots. Professor Martin[8] noticed a hoary powder on a Larch which grew near his house, in the year 1798, but neglected to examine whether it had any of the qualities of manna. The following year there was no appearance of this powder, which (as the Professor observes), may be accounted for from the latter being a cold, wet season, and the former a hot, dry one.

We are informed by the French author above quoted that the manna is collected by the peasants, who go

[1] Duhamel *Traité des Arbres*, Tom. l. p. 335.   [6] Duplessy, Tom. 2. p. 268.   [3] Lib. 16. c. 10.   [7] Pros. Med. Lib. 14. c. 1.
[2] *Mat. Med.*   [5] Pallas. *Flora Rossica*. Vol. I. p. 9.   " " *De Arb. coniferis.*" (Paris, 4to. 1653.) p. 9.   [8] Miller's Dict. Pinus Laris.

2 T

very early in the morning to the forests, and lop off with hatchets the branches that bear it, carrying these afterwards to the shade, where they can collect the grains at their leisure. It is then sold to merchants, who fix a considerable price upon it at a distance from the place where it is found, and the Venetians have many different names for the varieties of it. If kept longer than a year, it is apt to lose its proper taste and to be spoiled.[*]

In Dauphiny this manna has been very generally employed by the apothecaries as a laxative, but it is said to possess not more than half the strength of the product of the Calabrian ash. The mode in which it is formed, however, deserves to be more fully inquired into, and it would be satisfactory to ascertain whether there be any difference between this saccharine matter and the *Mel Cedrinum*, *roscidum*, and *aerium*, of the ancients. Galen (in his third book on aliments) describes a mode of collecting the " *Ros Montis Libani* " very similar to that given by Bellonius respecting the manna of the Larch.

BOLETUS LARYCINUS.

AGARICUM. *Bellon. Axtius*, tab. 20, 21. *Dale*, &c. *Jacquin Miscel. Austr.* 1. p. 164.
" *Boletus abies Laricis dicta.*" *Linn. Mat. Med.* 497.
Purging Agaric, (of the shops).

On the trunk of the Larch is sometimes found a remarkable species of *Boletus*, well known to botanists, and which is well described by Jacquin and Pallas.[*] It does not occur very frequently in the more southern parts of Europe, if we may credit Bellonius, but it is common in some parts of the Russian empire, and exported largely from Archangel. The form is generally oblong, variously lobed, and the lower part is always somewhat truncated, porous, and of a sort of mud colour, whilst the remaining surface has an ashen hue. Its *parenchyma* is soft, sweetish, (at last of a nauseous bitter taste) and saponaceous, whence it is used among the women in some parts of Siberia for washing their skins, and even their linen.

Bellonius[*] speaks of this fungus yielding a fine purple dye, which has been found out, it appears, by the Tungoos, these people employing a decoction of it with the roots of *Galium* for staining the hair of the rein deer, to ornament their persons.

As a medicine, the *Boletus Larycinus* is now very rarely employed in England, but it retains a place among the domestic remedies of the Russians, as an emetic in intermitting fevers, and for some female complaints. The Baschkirs sprinkle the powder on foul ulcers of cattle, as a detergent and antiverminous remedy.[*]

If it be intended to apply the *Boletus Larignus* to medicinal purposes, some caution seems to be necessary, in regard to the time of gathering it. Bellonius[*] recommends the autumn, that being the season at which, in common with fruits, he supposes this vegetable production to be just in a state of maturity. Should it have exceeded two years' growth, its qualities, he says, will have undergone a change of a deleterious nature; and if it should not have completed one year, the exhibition of it may be followed by effects equally pernicious.

[*] Vide locum supra citatum.   [*] *Fl. Ross.* Tom. 1. p. 3.   [*] " *De Arb. Conif.*" p. 26.   [*] Pallas. *(loco supra cit.)*   [*] p. 26.

COPY OF A LETTER

FROM

MR. THOMAS DAVIS, OF HOMMINGSHAM, WILTS,

RELATIVE TO THE

TIMBER YIELDED BY VARIOUS SPECIES OF PINES.

DEAR SIR,

HOMMINGSHAM, SEPT. 9, 1797.

I AM convinced, from repeated observations, that the Scotch Fir produces the deal called in London "*Yellow Deal*," and in the country "*Red Deal*," and being generally imported from Christiana, sometimes called "*Christiana Deal*." They frequently come hither in planks, but oftener in boards, *called* twelve inches wide, though seldom above ten inches and a half, cut through and through, or as the sawyers call it, cut "*fletch*." Of course the trees are not above twelve inches diameter, and yet I have counted their rings and found their growth to be from sixty to an hundred years. They must therefore grow thick together, and upon poor or rocky land, and this is also evident by the smallness of the knots, proceeding from the want of room to push out strong boughs.

The Scotch Fir raised in England is equal to the foreign in weight and durability, but is seldom so fine in the grain, and has a greater quantity of sap owing to its rapid growth, occasioned either by the superior strength of the land, or greater distance from one another, or both. But the quality is sufficiently similar to ascertain that they are the same species.

A foot square of Scotch Fir, English grown, and moderately dry, will weigh fifty-one; a foot of oak not much more than sixty-one.

A tree of an hundred years old (I look upon the ultimatum of its growth in England as not more than an hundred and fifty) may measure four load, or two hundred feet, and is fairly worth fifteen pounds.

Land planted with Scotch Firs eight feet and a quarter apart, viz. six hundred and forty to an acre, will pay ten per cent. *compound interest*, supposing very poor land at three shillings per acre, worth about four pounds in fee, and the planting to cost six pounds more, in all ten pounds per acre. In twenty-eight years ten pounds, at compound interest, will be forty pounds, and *in that period the trees at only two and sixpence each will be worth eighty pounds.*

Spruce Firs from which the deal we usually call *white* deal in England, is produced, are perhaps the next valuable to Scotch Fir; and, what is remarkable, those grown in England, are superior to any imported. That kind of tree not being hurt by knots, is the better for rapid growth, and the deal the handsomer. But it does not grow well in exposed situations. It there loses all its side branches, and not growing from leaders as a Scotch Fir does, gets mossy, lingers and dies: and if put close together, it never rises to any size. Perhaps it may be two or three load in one hundred years, worth seven or eight pounds, but an hundred years seem to be the full *ultimatum* of its growth. It is in fact fit for nothing but a garden, where *it is a pretty thing* for twenty or thirty years, when it grows naked, and should be removed and replaced by others.

The Silver Fir, the most beautiful in external appearance of all the genus, *either young or old,* grows much faster than either the Scotch or Spruce. At one hundred years old it is frequently above an hundred feet high, twelve or thirteen feet round, and contains at least six loads of timber, worth about fifteen pounds. The timber is more open, or (as the sawyers call it) "roacher in its grain," than the spruce, occasioned partly by the superior luxuriance of its growth, and therefore *should be used in large scantlings,* where its strength and toughness render it a valuable wood, particularly for beams; only great care must be taken that the ends are dry and accessible to air.

The Weymouth Pine is a *white pine,* but *still lighter and roacher* in the grain than the preceding sorts. Its principal use in its own country (America) is for masts of ships, for which its toughness makes it proper. It will, if placed in strong land well sheltered, get to four or five load in an hundred years, worth eight or ten pounds.

Larch is a delicate coloured wood, not unlike the Cedar used for black lead pencils, either in colour or smell. It has but little sap, and is convertible to flooring board at an early age, but its knots are then rather unsightly. We have few in this kingdom of a large size, and I have observed they decay and become mossy about forty years old. They grow best in sheltered situations. However it is a valuable and pretty looking wood, either standing or converted.

2 U

It is a mistake to suppose that fir trees should be cut in summer, because, (as they say) the sap, which is the turpentine, is afloat; they should always be cut when the sap is stagnant, *viz.* in winter. Fir cut in the summer will become full of mushrooms in a twelvemonth afterwards. I have tried this frequently and paid dearly for my experience.

PLACES OF GROWTH.

1. Scotch Fir. Mountainous rocky situations, shelter not necessary.
2. Spruce. Gardens and lawns, where it can have room to spread its lower branches horizontally, and is not wanted to stand above forty years.
3. Silver Fir. Strong lands, fit for oak; rather sheltered; but shelter not indispensable.
4. Larch. Sheltered situations on the sides of hills; sandy, if possible, though poor. It does not like cold wet land.

I omit the Balm of Gilead Fir, as we have had none grown to an age sufficient for determining its rise as timber. It seems to be very like the Silver Fir in quality.

I hope, Sir, you will find something in these remarks worthy your notice, as they are the result of the experience of above thirty years, *from the seed to the great tree.*

I am, Sir, your very humble servant,

THOMAS DAVIS.

EXTRACT

FROM

THE REV. WILLIAM COXE'S TRAVELS IN POLAND, RUSSIA, SWEDEN, AND DENMARK,

ON

THE SUBJECT OF CHRISTIANA DEAL,

(Vol. V. p. 28, 5th Edit.)

"The planks and deals are of superior estimation to those sent from America, Russia, and from the different parts of the Baltic, because the trees grow on the rocks, and are therefore firmer, more compact, and less liable to rot than the others, which chiefly shoot from a sandy or loamy soil. The plans are either red or white fir, or pine. The red wood is produced from the Scotch Fir, and the white wood, which is in such high estimation, from the Spruce Fir. This wood is the most demanded, because no country produces it in such quantities as this part of Norway. Each tree yields three pieces of timber, eleven or twelve feet in length, and is usually sawed into three planks; a tree generally requires seventy or eighty years growth before it arrives at the greatest perfection.

The environs of Christiana not yielding sufficient planks for exportation, the greater part of the timber is hewn in the inland country, and floated down the rivers and cataracts. Saw-mills are used for the purpose of cutting the planks, but must be privileged, and can only cut a certain quantity. The proprietors are bound to declare on oath, that they have not exceeded that quantity; and if they do, the privilege is taken away, and the saw-mill destroyed. There are one hundred and thirty-six privileged saw-mills at Christiana, of which one hundred belong to the family of Anker. The quantity of planks permitted to be cut amounts to 20,000,000 standard deals, twelve feet long, and one inch and a quarter thick."

In Scotland, they distinguish the wood cut in the native forests from that obtained in plantations, by calling the former *Highland Fir,* and the latter *Park Fir.* The Highland Fir is most esteemed, on account of its greater durability, being frequently found undecayed in ancient buildings, when the other sort is entirely wasted. This striking difference in the same species is probably to be attributed to the mountainous and rocky situations in which the native timber is found, and where the trees being of slower growth the wood is consequently of a harder texture; the latter may be readily distinguished from that of the Park Fir by its much deeper yellow colour.

COPY OF A LETTER

THOMAS MARSHAM, ESQ. TR. L. S. TO THE AUTHOR,

ON THE

SUBJECT OF INSECTS DESTRUCTIVE TO PINES.

DEAR SIR,
BAKER STREET, JUNE 6, 1803.

AGREEABLY to your desire, I have perused with care and attention the several letters from your friends in Wiltshire respecting the insect which attacks the various species of Pine trees in that county. It is evidently the Dermestes piniperda of Linnæus, (Ips piniperda of Coleoptera Britannica and De Geer) and although apparently so common and destructive, is yet but little known in its larva or grub state, for all the observations seem to relate to the perfect insect, whereas most animals of this kind are, in general, found to be far more mischievous in their primary or larva state. Linnæus observes that it perforates the inferior branches of the Pine; De Geer remarks, that he has found it in the wood of weak trees, and also within the young green branches, which it hollows interiorly by eating the substance, and this causes the branches to dry and perish; but neither of those authors mentions the particular species of Pine to which it gives a preference. Mr. Wickham in his letter to you states that he found it on Pinus sylvestris, Pinea, and Strobus, but that its greatest ravage was on Pinus sylvestris; that he had not perceived it on Pinus Cembra, abies, or on either species of Larix. Mr. Davies observes, " that it bores a hole through the shoots of the " last spring about eight inches from the summit, and then works its way up the pith, whereby the branch " withers and breaks off; and as it attacks not only the main leader, but also the side leaders, of course the " tree will lose all its leaves the second year, and must inevitably die." The devastation mentioned by the Duke of Somerset and Lord Malmsbury, which had taken place in the Pine forests in Germany, I am of opinion cannot be attributed to the Ips piniperda, but to the larva of the Phalæna Bombyx Monacha of Linnæus, which consumes the leaves of the Pine, and which you as well as myself have been lately informed has destroyed whole forests thirty miles in extent. D. Johann. Heinrich Jordens, in a work published in German in the year 1798, says, that the larva of this moth has discovered its mischievous tendency these two years, by entirely destroying the forests of Schlier and Ebendorf, and has now begun to spread itself on the confines of Voightland towards Bayreuth, where it attacked in a circumference of eight to ten German miles (i. e. forty-eight to sixty English) several larger and smaller forests. In 1784, he observes, that in the Selber forest it eat up the Fir as well as the Pine trees; but this circumstance had not been noticed in general, and he thinks that nothing but extreme hunger could force them to attack the needles of the Fir tree. The larva of the Tenthredo Pini is another very pernicious little animal; a few of them were sent some years since from Scotland, which I had to examine: they had destroyed an immense and valuable plantation of Pines. This latter insect is very accurately described by that celebrated Swedish naturalist the Baron De Geer. The Curculio Pini and Curculio Abietis are said to be very prejudicial to trees of the Genus Pinus, but as the larvæ of these, I presume, feed on the substance of the wood, their larvæ are but little known. Six other insects are also enumerated by the German writers as destructive to the Pine, viz. Dermestes Typographus, Polygraphus, Micographus, and Calcographus, Dermestes Scolytus, and Cerambyx Inquisitor; to which a much longer list might be added, both of those that feed on the leaves and those that penetrate into the trunk itself.

We have most, if not the whole, of these insects in our own country, but happily we hear but of little mischief occasioned by them, and they are but rarely found. Of Dermestes Typographus I have never yet found a single specimen; and of Phalæna monacha but one, during the number of years that I have made this branch of science the amusement of my leisure hours. The Dermestes Scolytus seems with us to confine its ravages to the Elm, of which mischief I have been lately an eye witness. I have thus endeavoured to comply with

2 x

your wishes in stating a few facts, which I fear will afford you but little information, nor do I by any means think them deserving publication, as they lead to nothing, unless that, as a spirit of inquiry seems to be taking place, they may act as a stimulus to engage gentlemen who have leisure, and possess large plantations of Pines, to examine into and investigate the nature, habits, and œconomy of these small destructive beings. And in that case I would recommend to their particular observation, to attend to and watch their progress, to search for and discover their natural enemies, and endeavour to employ them one against the other, for each of them is food for some other animal, which if encouraged will do more towards extirpating them, than all the powers which human skill could effect: for I am fully convinced that we often destroy the friend and spare the foe, for want of proper knowledge and discrimination; and this frequently in the case of birds, who are very great destroyers of insects. I would advise persons whose interest is concerned in saving their plantations and crops, whenever they shoot a bird, carefully to examine its craw, and note what it contains, by which means they would soon gain valuable information, and know when to destroy and when to encourage. The great desideratum of the Entomologist is to know his insect in all its stages, its habits, and œconomy; and for this they must require the assistance of persons resident in the country, who have these things constantly before their eyes, but who pass by them without the slightest notice, or even consider them as deserving attention, though perhaps they are suffering most materially.

I can only add, that if this hasty production gives you any pleasure or satisfaction, the end is answered to,

Dear Sir, yours truly,

THOMAS MARSHAM.

A. B. LAMBERT, Esq.

THE END.

T. BENSLEY, Printer, Bolt Court, Fleet Street, London.

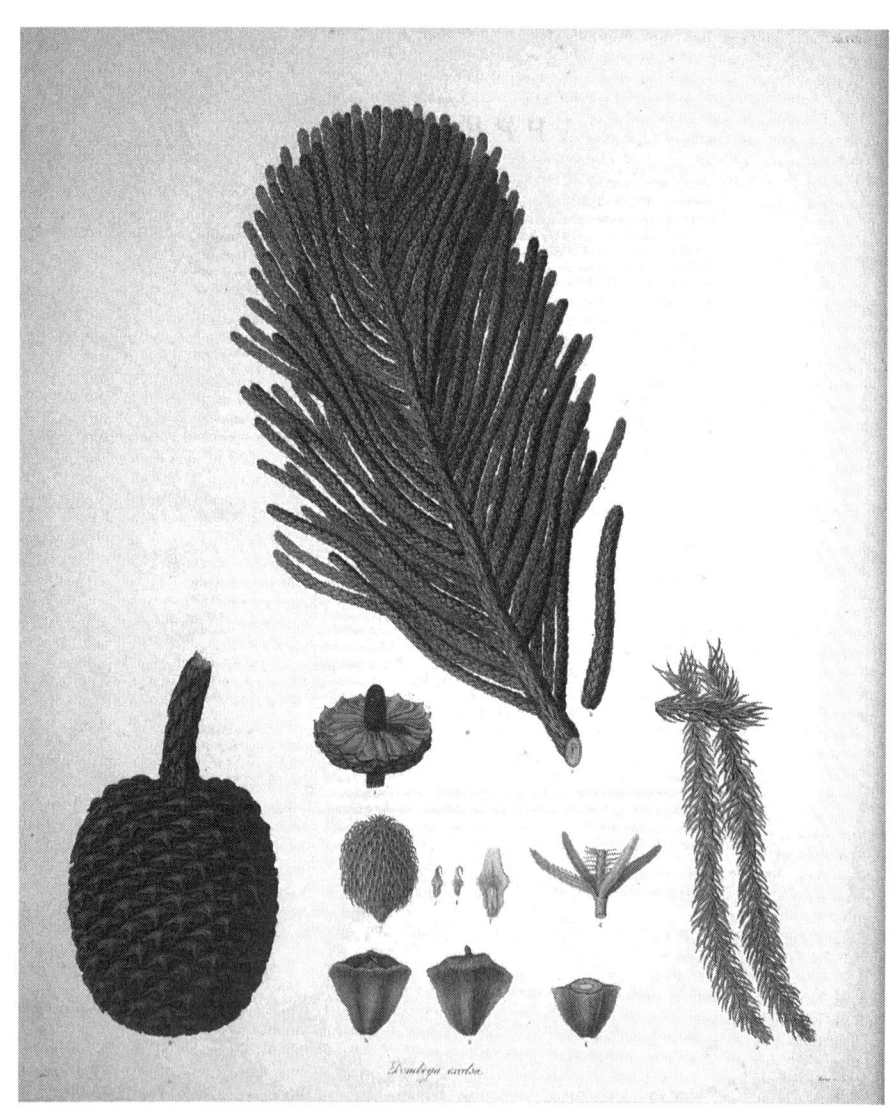

Dombeya excelsa.

APPENDIX.

DOMBEYA.

DIOECIA MONADELPHIA. *Linn.* CONIFERÆ. *Juss.*

Flores amentacei dioici.

MASCULI.

CALYX. *Amenti* squama, antheris utrinque tecta.
COROLLA nulla.
STAMINA. *Filamenta* nulla. *Antheræ* squamis adnatæ, oblongæ, numerosæ. *Lamarck.*

FŒMINEI.

CALYX. *Strobilus* constans squamis imbricatis, unifloris, persistentibus, induratis, singulis *stigma*
ferentibus.
COROLLA nulla.
PISTILLUM.
SEMEN solitarium, intra basin squamæ.

TAB. 39 & 40.

DOMBEYA EXCELSA.

NORFOLK ISLAND PINE.

DOMBEYA EXCELSA, foliis adultioribus arctè imbricatis inflexis muticis.
Cupressus *columnaris*, foliis imbricatis subulatis sulcatis, strobilis cylindricis elongatis. *Forst. Flor.*
Ins. Aust. 67.

THIS tree, which forms a new genus, very distinct from that of *Pinus*, is the tallest at present known, being, according to Governor Phillip, one hundred and sixty, or even one hundred and eighty feet in height; and Governor King informs us he measured some in Norfolk Island which were two hundred and twenty eight feet high, and eleven in diameter! It was first discovered by Captain Cook in his second voyage round the globe, on that extremity of New Caledonia, called Queen Charlotte's Foreland, and on a small neighbouring island, named by Captain Cook Botany Island, which is a mere sand bank, only three-fourths of a mile in circuit; also on another island, called by our voyagers the Isle of Pines, from its being almost covered with the above-mentioned tree. A fine view of this may be seen in Cook's second voyage, vol. ii. 140. representing several of these very singular trees. His words are, " If I except New Zealand, I, at this time, knew of no island in the South Pacific Ocean, where a ship could supply herself with a mast or a yard, were she ever so much distressed for want of one. Thus far the discovery is or may be valuable. My carpenter was of opinion that these trees would make exceedingly good masts. The wood is white, close grained, tough and light. Turpentine had exuded out of most of the trees, and the sun had inspissated it into a rosin, which was found sticking to the trunks, and lying about the roots. These trees shoot out their branches like all other pines; with this difference, that the branches of these are much smaller and shorter; so that the knots become nothing when the tree is wrought for use. I took notice that the largest of them had the smallest and shortest branches, and were crowned as it were, at the top, by a spreading branch like a bush. This was what led some on board into the extravagant notion of their being basaltes." Forster, speaking of them in his account of the above voyage, says, " Peculiar to Norfolk Island, and to the east end of New Caledonia, we found a species of coniferous tree, from the cones seeming probably to be a cypress; it grows here to a great size, and is very heavy, but useful timber.".

In Captain Hunter's *Journal of the Transactions at Port Jackson and Norfolk Island*, p. 194, we have also the following excellent account. " The Pines, which have been particularly spoken of by Captain Cook, and by others, who have lately visited this island, are the most conspicuous of any trees here; they grow to a prodigious size, and are proportionably tall, being from one hundred and fifty to two hundred feet, and in circumference from twelve to fourteen feet, some to twenty-eight and thirty feet. These trees, from their immense height, have a very noble appearance, being in general very straight, and free from branches, to forty, sometimes sixty feet, above the ground; they have been by some thought fit for masts, for ships of any size; in length and diameter they certainly are, but with respect to quality they are, in my opinion, wholly unfit; even admitting them to be sound, which, from experience, I know is seldom the case. I employed the carpenters of the Sirius, while here, to cut down a few sticks, which it was intended should be sent home by the first opportunity, in order for trial in his Majesty's dock-yards, to see if they were, as had been said, fit for his Majesty's navy, or not. In providing a top-mast and a top-sail-yard for a seventy-four gun ship, a thirty-two, a twenty, or a sloop, and one rough spar, in all seven sticks, thirty-four trees were cut down, twenty-seven of which were found defective. When these trees were falling, it was observed that most of them discharged a considerable quantity of clear water, which continued to flow at every fresh cut of the axe; there is no turpentine in these trees but what circulates between the bark and body of the tree, and which is soluble in water. It is a very short grained and spongy kind of timber, and I think fit only for house-building, for which we know it to be very useful. When fresh cut down, five out of six will sink in water, the wood is so exceedingly heavy: and, if we suppose for a moment, that great part of the pine timber were fit for naval purposes, the great difficulty, and indeed I may say impossibility, of getting it from the interior parts of the island to the sea, would render it of little value, if designed for masts; but if for plank, it could be cut up where fallen. Those which grow on the south-east point of the island, where the land is low, are those which have hitherto been made use of."

And at p. 389 of the same work, is another very interesting account from the journal of Lieutenant (now Governor) King: he there says, " The pine-trees are of a great size, many of them being from one hundred and eighty to two hundred and twenty feet high, and from four to eight feet diameter some distance from the ground. Those trees, which measure from one hundred to one hundred and eighty feet high, are in general sound, and are without branches for eighty or ninety feet, but the upper part is too knotty and hard to be useful; indeed, it frequently happens, that after twenty feet have been cut off from the butt, the trees become rotten and shaky, and are also very brittle; for which reason, no dependance can be put on them for masts or yards. The turpentine which exudes freely from the bark, is of a milk-white glutinous substance; but it is rather remarkable, that there is none in the timber. We tried to render this turpentine useful in paying boats, and other purposes, but without success; as it would neither melt nor burn: we also tried to make pitch or tar, by burning the old pines; but there being no turpentine in the wood, our efforts were useless. The pine is very useful in buildings, and being dispersed in various parts of the island, is well calculated for such buildings as hereafter may be necessary: from what I have been able to observe, it is very durable, as that which we had used for erecting houses, stood the weather very well."

That excellent botanist Mr. Brown, who was on board the Investigator with Captain Flinders, in his late voyage of discovery round New Holland, informs me, that they found this tree growing in great abundance on several parts of the east coast; and that he climbed several of them, but could not find any fructification: those he saw were not above sixty or seventy feet in height. A bay on the above-mentioned coast, from the great abundance of these trees found there, they named the Bay of Pines. A beautiful drawing of this spot, by Mr. William Westall, landscape painter on board the above-mentioned ship, was in the Exhibition of 1805. A few of these trees are now in the gardens of the curious about London; they thrive exceedingly well in our green-houses, grow very fast, and are one of the greatest ornaments of our collections. They can be increased by cuttings, but with great difficulty; and never by this method make handsome plants. A most beautiful specimen, and the largest in this kingdom, is now in the Royal Gardens at Kew, to which it was first introduced by Governor Phillip. Fig. a. some Botanists have suggested might belong to another species, because the branches are somewhat smaller than others that have been brought from Norfolk Island; but this, can now no longer remain a doubt, since that place has been so well explored by that accurate botanical draughtsman, Mr. Ferdinand Bauer.

I must here observe a remarkable peculiarity belonging to the *Coniferæ* of the southern hemisphere, which is, that while the trees are young their leaves are long and divaricating, but when they become old enough to bear fruit, those leaves fall off, and are succeeded by short scales closely imbricated on the branches, so that seeing them in their different states, one could hardly suppose it possible that they could belong to the same species. This is very remarkable in one of the species of *Dacrydium* from New Zealand, where the leaves, whilst the tree is young, have the appearance of the common yew, (*Taxus baccata*,) but become imbricated scales, and somewhat resembling *Juniperus Virginiana*, when the tree grows older.

TAB. 39.

a. Represents a branch in Sir Joseph Banks's Herbarium, brought home by Captain Cook, marked
"Nova Caledonia, Isle of Pines. W. Anderson, 1774."

b. Part of a branch from Norfolk Island.

c. Branch with leaves of the first growth, from the Kew Garden, the plant brought from Norfolk
Island.

d. Cotyledons (which are always four in number) from Norfolk Island.

e. A very young cone belonging to branch a.

f. Scales of the same cone natural size. F. scale magnified.

g. Ripe cone from Norfolk Island.

h. Scales separate.

i. Rachis of the cone, reduced to half its natural size, shewing the disposition of the scales.

k. Transverse section of a scale, shewing the seed.

TAB. 40.

A Branch of *Dombeya excelsa* from Norfolk Island

ADDENDA ET CORRIGENDA.

Tab. 16. " Having lately been favoured with some specimens of Pines, by William Strickland, Esq. of Yorkshire, who collected them in America, I am enabled to correct an error I have committed in the sixteenth plate of my work, where I have figured a cone of a new species, as belonging to Pinus Tæda. I was led into this from receiving it with the branch there figured from America, and understanding at that time that they were from the same tree. Having now received a branch belonging to the cone gathered by Mr. Strickland himself, the leaves of which are shorter and broader than those of Pinus Pumilio, and terminating in a very sharp spine, sufficiently distinguishing it from all other species, I shall call it, from the peculiar spines on the cones, PINUS pungens, *foliis geminis brevibus acutis, strobilis ovato-conicis; aculeis squamarum elongatis subulatis: superioribus incurvis: inferioribus recurvis.*

A figure of a branch, with male flowers, I hope to be able to give at some future period. Mr. Strickland found large forests of this Pine on the summit of the Blue Mountains, on the Frontiers of Virginia and N. Carolina.

From the cones he brought home he was only able to raise one tree, which is now growing at his seat in Yorkshire, above six feet high, and I have no doubt it is the only one at present in England." Annals of Botany, vol. ii. 198. *

In Tab. 21, the cone d, d. with its dissections are now given from authentic specimens brought home by Dr. Roxburgh, instead of those which were figured before by mistake, and which belong to another species not yet described; therefore in page 29, lines 13, 14, for " *ovati parùm incurvi*," &c, &c. read " *ovato-conici, squamarum apices elongati, obtusi, recurvi.*"

Tab. 34. *Pinus lanceolata*, has been coloured, and an additional branch given, from a fine drawing at the India House, sent to their Museum by the Company's Botanical Draughtsman at Canton.

Tab. 37, bis. *Pinus Cedrus*, taken from a fine tree in the Royal Gardens at Kew.

Tab. 38. *Pinus Dammara*. For the additional specimens now represented, I am obliged to Sir Joseph Banks, who received them from Amboyna. They were sent home in spirits by Mr. Christopher Smith. The plate is coloured from a plant growing in the Royal Garden at Kew. It will readily be perceived that this tree must form a new genus.

* Since the above was written, I have raised several young plants from the cones given me by Mr. Fraser.

Tab. XLI.

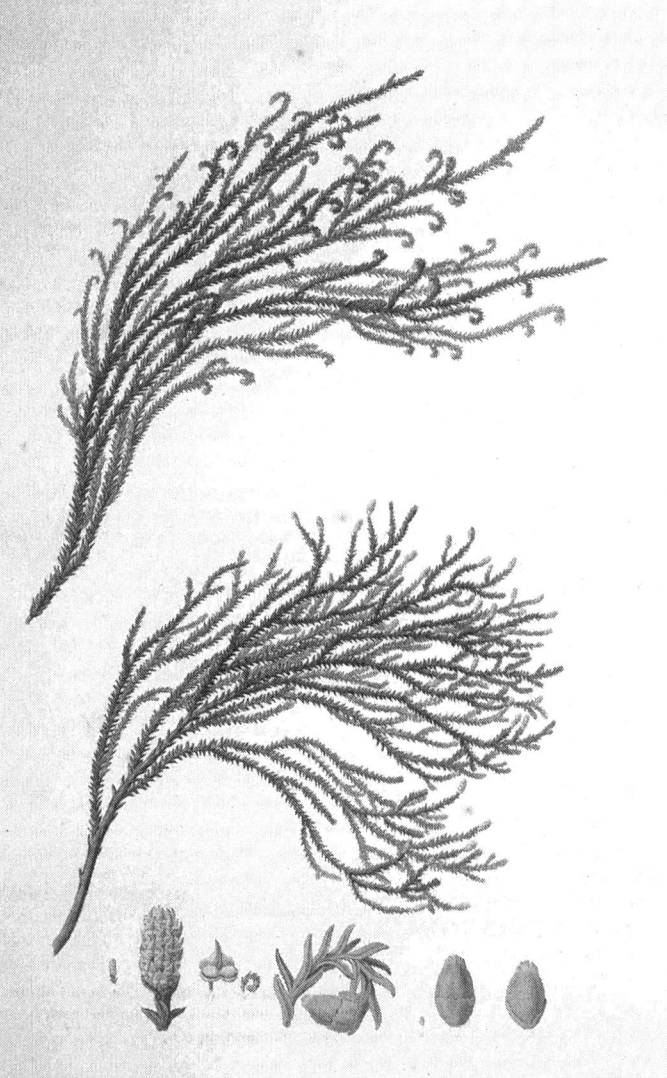

Dicranum scoparium.

DACRYDIUM.

DIOECIA MONADELPHIA. *Linn.* Coniferæ. *Juss.*

FLORES AMENTACEI DIOICI.

MASCULI.

CALYX. *Amentum* oblongum, constans squamis cordato-acuminatis, staminiferis.
COROLLA nulla.
STAMINA. *Filamenta* nulla. *Antheræ* didymæ, orbiculares, transversim dehiscentes, squamæ basi adnatæ.

FŒMINEI.

CALYX.
COROLLA.
PISTILLUM
SEMEN. *Nux* ovata, monosperma, receptaculo ampliato, firmo, basi immersa.

TAB. 41.

DACRYDIUM CUPRESSINUM.

NEW ZEELAND SPRUCE.

DACRYDIUM CUPRESSINUM. *Solander apud Forst. Pl. Escul.* 80. *et Flor. Inst. Aust. Prod.* 92.
The Spruce Fir of New Zeeland. *Cook's Second Voyage, vol.* 1. 70. *tab. No.* 51.

Habitat in Nova Zeelandia.

DESCRIPTIO.

Arbor vasta, pulcherrima. Rami patuli, valde ramulosi, ramuli flexuosi, dependentes; foliis junioribus, subulatis, divaricatis; adultioribus laxe imbricatis.

THIS tree, as I am informed by Mr. Archibald Menzies, forms large forests in the south-west parts of New Zeeland. Captain Vancouver, who cut down several of these trees to refit his vessel, found the timber solid and close-grained, very much resembling the Bermudas cedar.

Forster, *De plantis Esculentis Insularum Oceani Australis, p. 80,* gives the following account of this tree. " This beautiful Genus allied to *Taxus,* whose flowers I have not seen, received the name of *Dacrydium* from the illustrious Solander. It is found in New Zeeland, where Cook, from the younger branches giving out a bitter resinous juice, prepared a liquor similar to beer, called Spruce, and like that beverage, also excellent in scorbutic disorders; but, drank on an empty stomach, it produces nausea and giddiness, which however go off in a short time."

The same author in his relation of Captain Cook's Voyage, page 130, also thus speaks of it. " The Spruce of New Zeeland is a very beautiful tree, and conspicuous on account of its pendent branches, which are loaded with numerous long thread-like leaves, of a vivid green. It frequently grows to the height of fifty or sixty, and even one hundred feet, and has above ten feet in girth."

Three species have been discovered in New Zeeland, which are among the unpublished figures of the Right Hon. Sir Joseph Banks, one of which, called *Dacrydium taxifolium,* is mentioned in Captain Cook's first Voyage, by Hawksworth, vol. iii, p. 441, as " growing in swamps, remarkably tall and strait, thick enough to make masts for vessels of any size, and, if a judgment may be formed by the direction of its grain, very tough: this which, as has been before remarked, our carpenter thought to resemble the Pitch Pine, may probably be lightened by tapping, and it will then make the finest masts in the world: it has a leaf not unlike a yew, and bears berries in small bunches." I received fine specimens of the branches of this tree from Governor Phillip, who procured them from New Zeeland by a vessel which touched there from New South Wales, but without either flowers or fruit, so that I am unable at present to give a figure of this interesting tree.

Tab. 41 is taken from fine specimens of *D. cupressinum* collected by Mr. Menzies, who very obligingly communicated them to me, and was coloured by permission of Sir Joseph Banks, from a beautiful drawing made by Sydney Parkinson, in his celebrated voyage with Captain Cook.

EXPLANATION OF TAB. 41.

    a. Male Catkin.
    A. The same magnified.
    B. Scale with antheræ magnified.
    c. End of a branch, with fruit.
    C. The same, magnified.
    d. The fruit.
    D. The same, magnified.
    D. The same, with part of the receptacle cut away to shew the insertion.

Since writing the former part of this work, I have noticed one of the finest trees of Pinus Cembra in the kingdom, growing at Beckenham Park, the seat of John Cator, Esq. given to the late Mrs. Cator by the celebrated Peter Collinson her father.

TAB. CCLXV.

CUPRESSUS.

Monoecia Monadelphia. *Linn.* Coniferæ. *Juss.*

FLORES AMENTACEI MONOICI.

Masculi.

Calyx. *Amentum* ovatum, squamosum, squamis oppositis, pedunculatis, quadrifariam imbricatis, sub-peltatis, carinatis, acuminatis, inflexis.
Corolla nulla.
Stamina. *Filamenta* nulla. *Antheræ* quatuor oblongæ, uniloculares, basi squamarum adnatæ.

Foeminei.

Calyx. *Strobilus* depressus, foliis obvallatis, constans *squamis* lanceolatis, sessilibus.
Corolla nulla.
Pistillum. *Germina* plurima, subrotunda, compressa, stigmatibusque coronata, sessilia. *Styli* nulli. *Stigmata* prominentia, subcylindracea, truncata, apicibus concavis.
Strobilus squamosus, subglobosus, squamis quadrifariam oppositis, pedunculatis, peltatis vel angulatis, extrorsum convexis, mucronatis, induratis, coarctatis, demum dehiscentibus.
Semina oblongo-ovata, angulato-compressa, alata, pedunculis crassis squamarum (ad singulum 4-30) seriatim affixa, dehiscentibus squamis decidua.

CUPRESSUS LUSITANICA.

CEDAR OF GOA.

Cupressus Lusitanica, ramis flexuosis patentibus, ramulis quadrangulis, foliis quadrifariam imbricatis acutis carinatis glaucis adpressis.
C. *lusitanica* ramulis quadrangulis, foliis quadrifariam imbricatis adpressis glaucis carinatis, strobilis subglobosis, squamis mucronatis, ramis pendulis. *Willd. Sp. Pl. Tom.* 4. *p.* 511.
C. *glauca* foliis quadrifariam imbricatis, acutis; ramis patentissimis, infimis, subdependentibus. *Brotero Flora Lusit. Vol.* 1. *p.* 216.
C. *pendula* foliis imbricatis glandulosis, frondibus quadrangulis glaucis, ramis dependentibus. *L'Herit. Stirp. Nov. p.* 15. *t.* 8. *Hort. Kew. Vol.* 3. *p.* 373.
C. *glauca* foliis acutis glaucis glandulosis quadrifariam imbricatis, ramis dependentibus. *Lam. Enc. Meth. Vol.* 2. *p.* 243.
C. *lusitanica* foliis imbricatis apicibus aculeatis, ramis dependentibus. *Mill. Dict. No.* 3.
C. *lusitanica*, patula, fructu minori. *Tourn.* 587. *Duham. Arb. Vol.* 1. *p.* 198.

Habitat Goæ in India Orientali, nunc in Lusitania.
Floret Autumno.

DESCRIPTIO.

Arbor ramosa, procera, altitudine 50 pedum, et ultra. *Rami* sparsi, inordinati, flexuoso-patentes. *Ramuli* incurvati numerosissimi, junioribus quadrangalis, foliis undique tectis; adultioribus teretibus. *Folia* squamiformia, subamplexicaulia, basi lata, sursum attenuata, subulata, diu persistentia; *juniora* quadrifariam imbricata, glauca, dorso glandulâ resinosâ concavâ longitudinaliter exarata; *seniora* remotiuscula, vix imbricata, rigida, demum arida, rufa. *Amenta mascula* numerosa, ovata, obtusè octangula, terminalia, solitaria, e viridi-lutea, lineas fere duas longa: Squamæ circiter 20 convexo-concavæ, luteæ, apice extus subviridi. *Amenta fœminea* solitaria, foliis circumcincta, depressa, minuta. *Strobilus* ovato-globosus, rugosulus, muricatus, magnitudine Baccæ *Pruni spinosæ*, farinâ cæsiâ conspersus. *Squamæ* octo angulatæ, mucronibus elongatis, reflexis. *Semina* gilva.

THE branch represented in Plate 42, was from a tree growing in my conservatory, where it produced some hundreds of cones when not more than twelve feet high, the air of Wiltshire being much too cold for it unless protected in winter; yet I have seen it at the Marquis of Blandford's, at White Knights, flourishing all the year without any shelter, except that of other trees keeping the wind from it, and thriving equally well as with me at Boyton when under cover. Although we are informed in the Hortus Kewensis, from Ray's Letters, that this tree was cultivated in England so long ago as the the year 1683, yet it is still very scarce; nor have I met with it any where but at White Knights, where there are several young trees of it now in a very flourishing state; but as Mr. Lee at Hammersmith, of whom I procured mine some years ago, had at that time several more for sale, I should suppose that it must be in other collections. Brotero informs us, that in Portugal this tree grows much faster than the common cypress (C. sempervirens), but is of shorter duration, and its timber of a softer texture. Miller's account of this species is as follows: " The Third sort (C. lusitanica) is at present pretty rare in the English gardens, though of late years there have been many plants raised here, but this sort is not quite so hardy, I fear, as the common cypress, for the plants are frequently killed, or greatly injured, in severe winters, and in the hard frost in 1740, there was a large tree of this kind entirely killed in the gardens of his Grace the Duke of Richmond, at Goodwood in Sussex, which had been growing there several years; and in the year 1762 many large trees were killed. There are great plenty of these trees growing at a place called Busaco, near Coimbra, in Portugal, where this tree is called the Cedar of Busaco; and there it grows to be a timber tree, so that from thence the seeds may be easily procured. This tree grows naturally at Goa, from whence it was first brought to Portugal, where it has succeeded and been propagated; formerly there were some trees of this sort growing in the Bishop of London's Garden, at Fulham, where it passed under the title of Cedar of Goa, by which it was sent from thence to the Leyden Garden with that name." *Miller's Dict. l. c.*

EXPLANATION OF TAB. 42.

A. Male Catkin magnified.
B. Scale of the same.
C. Female Catkin magnified.
D. Germen and stigma magnified.
e. Scale of a half grown cone with the seeds.
F. One of the seeds magnified.
g. Ripe Cone. h. Scale of the same. i. Seed.

Tab. XXXIII

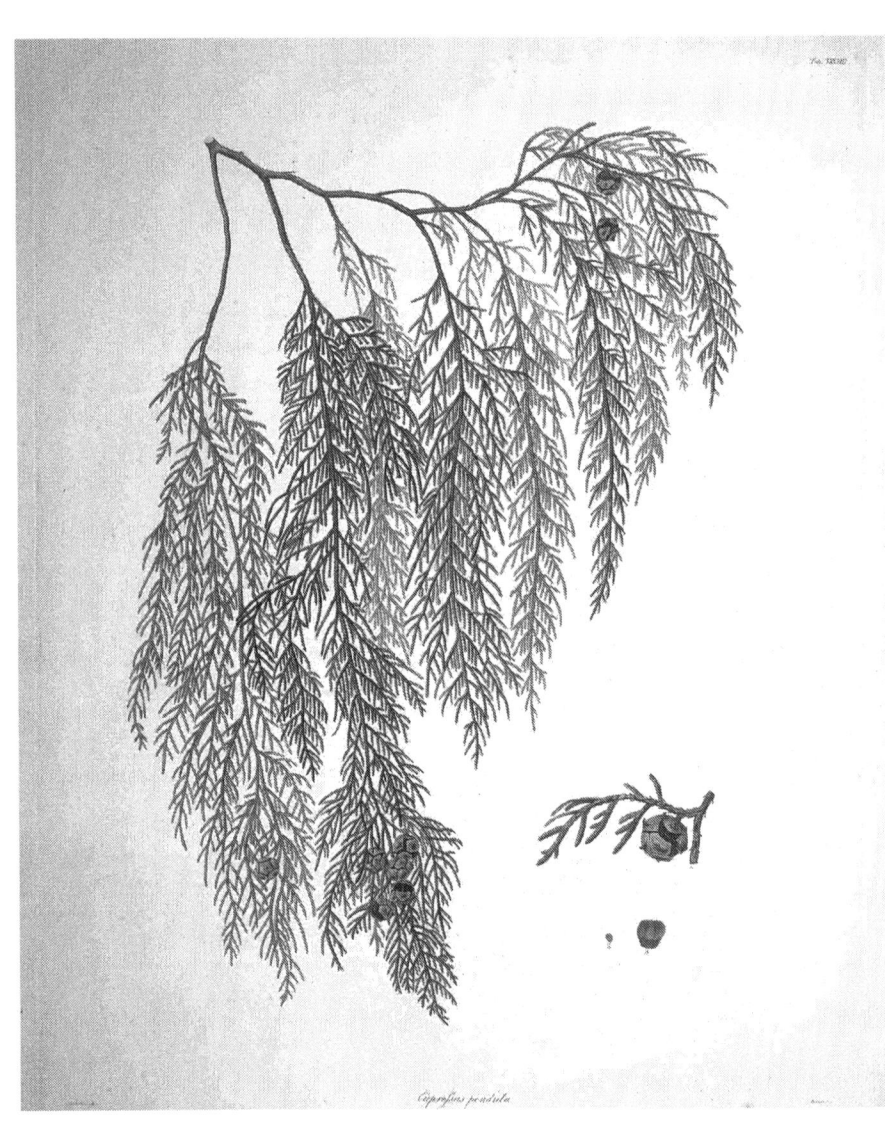

Cupressus pendula

CUPRESSUS PENDULA.

WEEPING CYPRESS.

Cupressus pendula ramulis ancipitibus foliosis, primariis longissimis pendulis, novissimis brevibus alternis, bifariam patentibus.

C. pendula foliis oppositis ovatis, ramulis dichotomis pendulis. *Thunb. Flora Jap. p. 265. Willd. Sp. Pl. Vol. 4. p. 512.*

C. pendula. *Staunton's authentic Account of an Embassy from the King of Great Britain to the Emperor of China, Vol. 2. p. 525.* Weeping Thuja, or lignum vitæ. *Idem, p. 445 et tab. 41.*

Fi-moro. Juniperus julifera, julis squamosis croceo polline refertis, baccis Sabinæ angulosis. *Kæmpf Amœn. Exot. p. 883.*

Habitat in China et Japonia.
Floret......

Arbor comâ amplâ, expansâ. *Rami* dichotomi, laxi, aphylli, valde ramulosi: Ramuli longi, compressi, penduli, foliis arctissime tecti, iterum divisi, secundariis brevibus, patulis. *Folia* quadrifariam imbricata, semiamplexicaulia, subtriquetra, carinata, adpressa. *Amenta* mascula numerosa, ovata, lineâ vix longiora, in apicibus ramulorum solitaria, sessilia: *fœminea* depressa, minuta, foliis patulis cincta, ramulos inferiores brevissimos terminantia. *Strobilus* fuscus, magnitudine fructus *Pruni spinosæ.* Squamæ octo angulatæ mucronibus obtusis. *Semina* gilva.

This elegant tree was discovered in Japan, and first made known to Europeans by the illustrious Kaempfer, and again found there, on the mountains of Fakonia, by Thunberg, who remarks, that its exceedingly long and numerous pendent branches give it a very remarkable and beautiful appearance, very different from the other ever-greens of the same natural order. This Naturalist, however, found but one tree, and neither saw the flowers nor the fruit.

In Plate 41 of the Account of the Embassy to China, in the View of the Vale of Tombs, it is represented as overhanging the monuments of departed greatness; nor, whether we consider its dark hue, or drooping pensile form, would it be easy to conceive any thing more in sympathy with the scene. No doubt it would also prove a great ornament to our gardens, into which it has not yet been introduced; but most probably we may soon be indebted for it to that skilful gardener, Mr. Ker, now employed in a mission to China for the purpose of collecting plants for the Royal Gardens at Kew, and who has already sent over a number of new and beautiful species from that interesting country.

The specimens represented in Tab. 43 were obligingly given to me by Sir George Staunton, Bart. who collected them while engaged in the celebrated Embassy to China. They agree so well with Thunberg's description of his *Cupressus pendula* above quoted, that, although I have not had an opportunity

EXPLANATION OF TAB. 43

A. Cone and branch magnified.
b. Seed natural size.
B. The same magnified.

of comparing them with each other, I have still considered them as the same species. Those possessing Japanese specimens will easily determine it from the figure.

Having in the first part of this work spoken of several gardens abounding in species of Pines, I shall here mention another which I have since become acquainted with, and have had several opportunities of visiting; a garden containing a greater number of species of the natural order of Coniferæ, perhaps, than any other in Europe: I allude to White Knights, the charming residence of the Marquis of Blandford, whose park and gardens are embellished with every species of hardy trees that can at present be procured; the number of species, and indeed, of individuals of each, is greater than that of any other collection I have had an opportunity of seeing. To form an idea of the scale on which these gardens are laid out, it need only be mentioned that one Arboretum occupied seven acres, and within the last two years has been considerably enlarged. The superior skill in cultivation of the noble owner, and its excellent soil and situation, have rendered the collection of hardy exotics at White Knights one of the most interesting which this country has produced.

BOYTON HOUSE,
Sept. 1807.

Note to PINUS HALEPENSIS, p. 16.

I have lately seen a most beautiful tree of *Pinus Halepensis*, bearing abundance of cones in the greatest perfection, in the garden of Joshua Smith, Esq. at Stokes Park, Wilts; it is by far the largest of the species I have ever seen, and the only one which I have found in fruit. Mrs. Smith informs me that it was planted by herself about seventeen years ago. The soil is sandy, and well sheltered by surrounding plantations.

FINIS.

Milton Keynes UK
Ingram Content Group UK Ltd.
UKHW022302280923
429602UK00005B/103